ghost
wave

GHOST WAVE

THE DISCOVERY
OF
CORTES BANK
AND THE BIGGEST
WAVE
ON EARTH

BY CHRIS DIXON

CHRONICLE BOOKS
SAN FRANCISCO

Dixon, Chris, 1966-
  Ghost wave : the discovery of Cortes bank and the biggest wave on earth / by Chris Dixon.
    p. cm.
  Includes bibliographical references.
  ISBN 978-0-8118-7628-5
  1. Surfing--California. 2. Surfing--Pacific Area. 3. Ocean waves--California. 4. Ocean waves--Pacific Area. I. Title.
  GV839.65.C2D59 2011
  797.3209794--dc22

                    2011020302

Manufactured in the United States

Designed by Jacob T. Gardner

10 9 8 7 6 5 4 3 2 1

Chronicle Books LLC
680 Second Street
San Francisco, California 94107
www.chroniclebooks.com

CONTENTS

ACKNOWLEDGMENTS

This book would not have been possible without the work, time, support, hospitality, and/or inspiration of the following people:

Will Allison, Grant "Twiggy" Baker, Rex Bank, Steve Barilotti, Rob Bender, George Beronius, Warren Blier, Daniel Martin Bresler, John Broder, Bruce Brown, Dana Brown, Rob Brown, Jimmy Buffett, Jeff Campbell, Steve Casimiro, Mike Castillo, Alfy Cater, Jeff Clark, Gary Clisby, Annouschka Collins, Josh Collins, Ken "Skindog" Collins, Sean Collins, Charles Coxe, Chris Crolley, Pat and Mary Curren, Don Curry, Jake Davi, Brett Davis, James Deckard, Jeff Divine, Jean Louise Dixon, Quinn Deckard Dixon, Richard Jobie Dixon, Watts Dixon, Shane Dorian, Dorothea Benton Frank, Lawrence Downes, Harrison Ealey, Grant Ellis, Dr. John English, William Finnegan, Nick Fox, Arthur "Mitch" Fraas, Matt George, Sam George, Brad Gerlach, Joe Gerlach, Dr. Gary Greene, Nancy Whitemarsh Gregos, Angie Gregos-Swaroop, Greg Grivetto, Nicole Gull, Jeff Hall, Laird Hamilton, Christine Hanley, Neil Hanson, Ellis T. Hardy, Christopher Havern, Steve Hawk, Mark Healey, Marty Hoffman, Philip "Flippy" Hoffman, Scott Hulet, George Hulse, Paul Hutton, Tom Jolly, Sebastian Junger, Dave Kalama, Ilima Kalama, Brian Keaulana, Buffalo Keaulana, Momi Keaulana, James Allen Knechtman, Eric Kozen, Dr. Rikk Kvitek, Randy Laine, Larry Kirshbaum, Steve Lawson, Adm. Robert J. Leuschner Jr., Brett Lickle, Kenneth Lifshitz, Brock Little, Greg Long, Rusty Long, Steve Long, Kate Lovemore, Gena and John Lovett, Leanne Lusk, Dr. Terry Maas, Don Mackay, Hugh MacRae Jr., Hugh MacRae Sr., Nick Madigan, Sarah Malarkey, Ben Marcus, Chris Mauro, Lucia McLeod, Garrett McNamara, Clement Meighan, Capt. Scott Meisel, Peter Mel, Tara Mel, Candace Moore, Larry "Flame" Moore, Dr. Walter Munk, Mickey Muñoz, Jason Murray, Ramon Navarro, Greg Noll, Laura Noll, Jeff Novak, Collin O'Neill, Dr. Bill O'Reilly, Dave Parmenter, Rebecca Parmer, Bob Parsons, Mike Parsons, Tara Parsons, Joel Patterson, Nate Perez, Steve Pezman, Judith Porcasi, Paul Porcasi, Jodi Pritchart, Mike Ramos, Rush Randle, Louis Ribeiro, Charles and Victoria Ricks, Anthony Ruffo, Roy Salis, Marcus Saunders, Bill "Dr. Evil" Sharp, Evan Slater, Kelly Slater, John Slider, Shari Smiley, Sunshine Smith, Kelly Sorensen, Jason Stallman, Capt. Steve Stampley, Jamie Sterling, Jean Stroman, Gloria Ricks Taylor *thanks mom!!!*, Kimball Taylor, Roy Taylor, Beverly Tetterton, Albert "Skip" Theberge, Brendon Thomas, James Thompson, Megan Thompson, Randy Thompson, Michele Titus, Matt Walker, Philip L. Walker, Les Walker, John Walla, Matt Warshaw, Grant Washburn, Frances "Taffy" Wells, Gerry Wheaton, James Whitemarsh, Brad Wieners, Malcolm Gault-Williams, Ben Wolfe, Matt Wybenga, Andrew Yatsko, Dr. Marvin Zuckerman

I would also like to thank:
My grandparents, for teaching me the value of a fine southern family and a damn good story.
My parents, for teaching me the difference between making a life and making a living.
Quinn, for teaching me the meaning of love.
Fritz and Lucy, for teaching me the meaning of life.

My sincere apologies to anyone I might have left out.

When foolhardiness would urge me to go and peep into some yawning chasm, my conscience would appear to say to me, "Stop! You are trifling with the Almighty!"

—A description of the first view of the caldera of Mount Kilauea, Hilo, Hawaii, September 1847, by Lieutenant Archibald MacRae, United States Navy (September 21, 1820–November 17, 1855)

Chapter 1:

THE
GHOST
WAVE

It was the only time I ever wrote
out a will before a surf trip.

—Bill Sharp

In the predawn hours of a dead-still December morning in 1990, a Black Watch sportfisher, its deck loaded with provisions, thick wetsuits, and big wave surfboards, motored out of Newport Harbor in Newport Beach, just south of Los Angeles.

Clearing the lights at the end of the harbor's long rock jetty, the skipper gave the boat's twin Yanmars their first big huff of diesel and crackling dry Santa Ana air. He then pointed the bow toward an empty spot, a big blank patch of ocean a hundred miles offshore where a ghost wave was said to appear, a wave of massive proportions that came out of nowhere, rose like a monster, and then slid back into the depths without a sign of its passing. According to legend, several vessels had met disaster here and now lay on the bottom, and the few mariners who had been out there told the surfers they were crazy. Along their intended route, compasses were known to spin in random directions. It was a place where the impossible was postulated to be an occasional nightmare reality—a breaking wave 100 feet high. They were headed for the Cortes Bank.

In addition to the captain, the boat contained four passengers: *Surfing* magazine editors Bill Sharp, Sam George, and Larry "Flame" Moore, along with a California pro surfer named George Hulse. Sharp, George, and Hulse were experienced big wave surfers, but in 1990, the world of monster swells was a far smaller and more mysterious place than it is today. The crucible of their sport still lay on Oahu at thundering tropical waves like Pipeline, Makaha, and Waimea Bay, and a relatively small group possessed the knowledge, skill, and guts to challenge them. Swell forecasting was still in its infancy; spots like Maverick's, Jaws, and Teahupoo lay far off the radar. Only recently, these three surfers had tested themselves on the first bona fide big wave find on the North American mainland—an icy, kelp-ringed giant off northern Baja's Todos Santos Island, appropriately named Killers. No one aboard had ever considered tying a water-ski rope to the stern of a Jet Ski and slingshotting a life-jacketed surfer onto a big wave—the pursuit today known as towsurfing. If you wanted to catch a big wave in 1990, you had to paddle like hell, pray, and never forget that if something went wrong, you were all alone.

Indeed, the surfers had gone to great lengths to ensure they were alone. This exploratory encounter with what they believed to be an unsurfed leviathan was the culmination of several years of painstaking, almost pathologically secretive detective work.

In December 1985, illuminated by the neon glow of a photographer's light table, Larry Moore pointed a freckled finger at page L4 of the *Chart Guide to Southern California*. "What about this spot? There's gotta be waves out there."

Beside him stood Sam George and Bill Sharp, the newly minted young editors of *Surfing* magazine. They had been scouring the nooks and crannies on the map, looking for places where they might find surf.

If there was one thing that George and Sharp had come to realize, it was that Flame was obsessive about everything he did. You didn't get a grain of sand in his Ford pickup. You didn't miss a 4 A.M. roll call for a photo shoot. You didn't mess with *any* element atop his photo desk. And you sure as hell didn't talk about surf spots you were scouting out. That was the great privilege and maddening frustration of the job. Larry possessed an obsessive need to know about the waves that broke along the Pacific Coast *and* to be the first to document them. Inclusion among his tight circle of explorers made you a very fortunate person, but you had to keep your mouth shut until Flame was ready to reveal a discovery—which might be never.

Flame was a fairly seasoned sailor. He had pored over his chart guides, studying coast and bathymetry from Vancouver Island to Cabo San Lucas. The same set of features that might sink a ship could also indicate a hidden wave. Lately, he had set his sights toward Todos Santos and San Clemente Island and now this weird shoal called Cortes Bank. He saw danger and opportunity. In fact, a mere month earlier, the *Los Angeles Times* had carried a story about the aircraft carrier USS *Enterprise* actually colliding with an unnamed reef "100 miles off San Diego." What other reef could it possibly be?

"Here's what it says," Flame read to Sharp and George. "Cortes Bank is about twenty-five miles long west-northwest to east-southeast by seven miles wide, with Bishop Rock awash and buoyed. The rock was struck by the clipper

Bishop in 1855 and is the farthest-outlying coastal danger. Nontidal currents of one to two knots cause much swell and moderate sea often breaks over the rock. A wreck near the rock is covered by only *three feet*. The bottom from five miles west-northwest to two and a half miles east-southeast is broken with hard white sand, broken shell, and fine coral. Anchorage is reported impractical due to swell, breakers, and lost anchors."

Sharp's blue eyes traced the tight contour lines. In addition to Bishop Rock, other shoal spots lay on Cortes Bank, one only nine fathoms deep. Another nine-fathom bank called Tanner lay just to the northeast. A few miles out, the ocean plunged to a thousand fathoms, or six thousand feet. "Good lord," Sharp said to Flame. "Three feet deep?"

Flame's first enquiry about Cortes Bank was with Philip "Flippy" Hoffman, a gruff old diver and local textile magnate. Hoffman had been among the very first Californians to challenge the giant, empty waves along the North Shore of Oahu in the early 1950s, and in 1973, he became one of the very first to surf Kaena Point, a frightful open-ocean wave off Oahu's easternmost edge. Hoffman moored his boat in Dana Point next to Flame's cherished *Candace Marie*, and he was as hard-core a waterman as you could ever hope to meet.

"I used to dive the Bank with the abalone fleet back in the 1950s, and I told Larry it had big wave potential," Hoffman said, his strong, old voice sounding as if it had been run through a fan.

"We'd go out there mostly in the fall. That's the nicest time of year for weather. I never saw it break all that big, and I never surfed out there because the currents are horrible and you couldn't stay in the lineup."

Diving was an isolated, dangerous business. Even with no breaking waves, the entire Bank was subject to tremendous, swirling surges of swell that could push or pull you sideways, or up and down in the water column, far faster than you might equalize the pressure in your ears. There were abalone the size of Bibles, lobsters the size of men, and sharks the size of busses. Were you swept from your boat, a current that suddenly rose to two or even four knots could make return utterly impossible.

Hoffman recalled being able to see the top of Bishop Rock, a pinnacle of hard volcanic basalt, in the trough of waves on a very low tide. "We went, maybe, four or five times from 1951 to 1958, just commercial fishing for abalone," Hoffman said. "The water could be very clear or dirty with plankton, and the fishing was just not quite as good as we thought it would be. It was a very rough place to try to sleep at night. Cups and plates would fly across the galley. I knew sometimes it must get really big out there."

Hoffman also told Flame that at least one diver—a famous Hawaiian big wave surfer named Ilima Kalama—had very nearly died out there.

In short, the Bank was not a place to be trifled with.

After that, Cortes Bank became an obsession for Flame. In January 1990, he and a gonzo surfer and bush pilot named Mike Castillo decided to see it for themselves. A now-legendary swell had just blitzed Hawaii, and they wanted to see what happened when it reached Bishop Rock, Cortes Bank's shallowest reach. From Castillo's tiny Cessna, a few hundred feet up in the air, Flame and Castillo were shocked to find a titanic, unruly wave unloading onto the submarine reef. Flame had traveled the world in search of surf, but he had never seen anything like this. A mile-long mutant Malibu was reeling off in the middle of the ocean. Castillo dove low and flew alongside a wave from a height of around thirty feet. Astonishingly, they appeared to be looking *up* at the wave's cascading lip.

"If anyone ever tries to surf out there," Castillo said, "they'd better take the fucking Pope along to pray for them."

A few days later, Flame showed photos of Cortes Bank to Sam George and Bill Sharp, swearing them to secrecy, as always. They were stunned. The photos raised disturbing and perplexing questions. Being big wave surfers, the most important was: How big was it? In the photographs, the only point of reference was a red marker buoy that disappeared in the maelstrom of white water at regular intervals as the waves passed. Flame didn't reveal the fact that he and Castillo had actually observed a wave from near sea level.

Bill Sharp recently mused, "If I knew how big the waves in those photos really were—or how big Flame thought they *might be*, I'm not sure if I would have gone. And if I had, it sure as hell wouldn't have been on such a tiny boat."

Eleven months later, and not long after exiting Newport Harbor in darkness, Sharp offered to take the helm of that tiny boat. He had a good basic understanding of LORAN navigation systems (GPS was not yet commonplace), and he was wide awake, so everyone else bundled up and went to sleep. The plan was to motor the twenty-nine-foot Black Watch for twenty miles out and around the southern end of Catalina Island. They would then cross another thirty-two miles of ocean to the southern flank of San Clemente Island, a naval base and artillery range populated by unexploded ordnance and a dwindling herd of feral, shell-shocked goats. From there, it was a simple, straight shot across forty miles of far wilder water. They had deliberately not notified the surfer-filled ranks of the Los Angeles or San Diego Coast Guard sectors of this expedition. "Loose lips sink ships," Flame told Sharp.

As the boat droned past Catalina, the first rays of sunlight painted the sky a pinkish purple. In the island's lee, a whisper of Santa Ana breeze carried the scent of chaparral and decaying bull kelp. Rising and falling over the butter smooth

Pacific, Sharp uneasily pondered the last-minute nature of this mission. Despite seeing photos, the surfers were essentially flying blind. Once the Black Watch cleared the shadow of San Clemente Island, the swell would become much bigger. But just how big? Sharp was particularly troubled by the rumor that this phantom wave had once scuttled a huge ship somewhere near the surf zone; a wreck was listed right there on the chart. What if some jagged piece of hull lay on the bottom? Getting stuffed by thirty feet of white water into a rusty portal was not a hazard most surf spots presented. What if the Santa Ana winds defied the forecast—as they often did—and wound up to hurricane force? What if a fog rolled in? Sharp thought of VW-size elephant seals and the creatures that dine on VW-size elephant seals. It was as if they were setting out to find and ride Moby Dick—bareback—and Sharp knew how that story ended.

The winds remained calm, but the undulating swells increased markedly. By the time the Black Watch rounded San Clemente Island three hours later, the crew began to stir, and Sharp informed his fellow surfers that they were dropping into the troughs of swells six to eight feet high at regular intervals of between seventeen and eighteen seconds. It was a solid west swell.

Sharp, Hulse, and George had all followed somewhat similar paths into the world of competitive surfing, but by 1990, none ranked at the top of the sport. Each began his career as a representative of the amateur National Scholastic Surfing Association's National Team. Hulse and George competed atop traditional surfboards, while the iconoclastic Sharp chose to ride his waves on a kneeboard. This kneebound surfing earned ribbing from Sharp's buddies, but that typically ceased when they saw him charge through suicidal barrels or launch himself onto waves on his short, stubby rocket that standup friends—including Sam—wouldn't touch. Sharp had developed a particularly fierce reputation at Todos Santos and at a mutant neck-breaker of a wave in Newport Beach called the Wedge.

Sharp was the son of a hard-charging Air Force fighter pilot. He had studied business at San Diego State University, where he founded the school's surf team. Hulse and George went on to compete in the ASP World Tour, a championship series of contests run by the nascent Association of Surfing Professionals. By 1989, Sharp and George had found their way—somewhat unexpectedly—into the small world of surf journalism, while Hulse, ground down by nonstop travel and a debaucherous party scene, had quit the World Tour. He was not nearly so widely known as Flame's "A-listers," pro surfers like Tom Curren, Brad Gerlach, Dave Parmenter, or budding West Coast big wave specialist Mike Parsons. Fortunately for Hulse, on this day all were off competing.

Sam George didn't share Sharp or Hulse's big wave bloodlust, but he could hold his own in most of the world's more radical lineups. He regarded the polished water and surging swell. "A lot bigger than it was when we left," he said to Sharp. "I wonder what the hell we're getting ourselves into."

"Shit, man," Sharp chirped, clutching the wheel and striking his best sea captain's pose. "Adventure is our business!"

At around 11 a.m., the LORAN indicated that the Black Watch was approaching the shallow southern periphery of Cortes Bank. "Something's going on," Sharp told George. "Look at the horizon."

Rather than the ruler-straight undulations of the previous several hours, the wave pulses suddenly steepened. They approached from odd angles, wobbling and lurching toward the boat like punch-drunk ski moguls. There was no obvious cause, but there was a reason.

The boat had rapidly passed from waters more than a mile in depth to the 150-foot-deep edge of a vast sunken mesa, which had disappeared beneath the Pacific Ocean a mere four thousand years ago. Swells whose energy columns ran nearly twelve hundred feet down were reacting to the first obstacle since slamming into the Hawaiian Islands twenty-five hundred miles ago. The Black Watch was built with a stable V-shaped hull, perfect for offshore fishing missions, yet she swooned from starboard to port. Confused, lumpy seas like this wouldn't have been all that unusual in a gale, but the air remained warm and calm. Sharp hoped the depth finder, which indicated that they were still motoring safely in more than a hundred feet of water, was functioning properly. He eased the throttle back a notch and strained for any point of reference. None was to be found.

Fifteen minutes later, the LORAN seemed to indicate that the Black Watch was still on a correct approach to Bishop Rock, but the team still saw no breaking waves. Had they entered an incorrect course heading? Was the swell too small?

"It's gotta be out here," Flame said, nervously staring through binoculars from the boat's upper platform like a sailor on the *Pequod*. "It's just gotta be."

Off the bow a few miles distant, weird ripples, a glint of sunlight, and a wisp of mist grabbed Sharp and George's attention. A surfacing whale? A gap in the swells gave a full view as another humpbacked shape breached in the same spot—followed by geysers of offshore spray. "It's a wave," Sharp yelled. "Thar she breaks!"

Flame began to unpack his camera gear, a flashbulb smile lighting the deepest creases of his face.

"It was just the most fantastic feeling," Sam George says today. "We had found Flame's Moby Dick."

Within a few miles, they spotted Bishop Rock's swaying warning buoy—Flame figured it was the same one he had seen from the air—and set a course that soon put them within earshot of what seemed the loneliest bell on the face of the planet. The tiny man-made island was laden with guano and inhabited by an argumentative posse of eight or nine sea lions. Sharp realized with shock that the buoy was big—maybe twenty-five feet tall. In the photographs, the white water from the broken waves completely buried the buoy, and thus must have been 40 to 50 feet high—bigger than any Flame had ever photographed. Yet that meant when the waves first broke, they would have been perhaps twice that high—bigger than anything anyone on board had ever imagined. What *was* this place?

Sets of waves appeared to the northeast of the buoy. Sharp approached the edge of the apparent surf zone on pins and needles. "We came up real slow," he says today. "We had no idea if there would be a rogue wave that might take us out, and so we just putted around for a while and watched. It wasn't really booming, the sets came every five or ten minutes. But when we finally got close and one rolled through, we were like—whoa, that's a rideable wave!"

The breaking waves were glacier blue. Silhouetted against the sky, the mist in their wake lit up like a million tiny shards of rainbow ice. Most of the waves weren't terribly steep, but they carried a great quantity of watery energy and seemed to approach the Bank at a terrific speed. They rolled, warbled, and peeled for a while and then disappeared back into the deep, continuing their march toward the coast of California. When a bigger one ran over what was obviously a very shallow spot on the reef, it reared up to vertical and threw out a beautiful, almond-eyed barrel. The surfers agreed that they seemed to resemble a cross between Oahu's Sunset Beach and Pinballs, a righthander that breaks along the inside of Waimea Bay.

This was going to be an exceedingly difficult place to surf. Every other wave they had ever ridden offered land-based points of reference—a hilltop, a dune, a palm tree, a lighthouse—some landmark that allowed a mental triangulation of position. Out here, it would not only be impossible to figure out where to sit in the water, but the featureless expanse greatly limited depth perception—making it impossible to judge the wave's size. Find yourself in the wrong spot, and you might be steamrolled and tumbled until you drowned or slammed down onto some nasty pinnacle of reef.

Hulse remembers, "You just had nothing to tell you where to be or how big the waves were. I was asking myself, is that 30 feet? Should I be writing out my will, too?"

To everyone's amazement Bill Sharp produced a bundle of bamboo poles, gallon plastic jugs, dayglo duct tape, and lead fishing weights from the hold and ordered the boys to get to work. "It was ingenious," says Sam George. "We were going to set a series of our little homegrown buoys to help triangulate a lineup."

Today, the surfers have forgotten the name of the Black Watch's wide-eyed, newly minted skipper, but now they handed him the helm, and Sharp and Flame worked mightily to convince him to reverse into position in the surf zone so the buoys could be laid. Backing in would allow for a fast forward escape should a set of waves lunge in from the deep at twenty-five knots.

As Led Zeppelin's *Kashmir* played over the stereo speakers, the team made fine work of tying the knots for the buoys. But when the roiling boat began to reverse, they inhaled greasy lungfuls of diesel smoke. Suddenly George began to feel queasy. "I thought, *What's wrong with me?*" he says. "Then it hit me. Oh my God, I'm getting seasick."

George ran belowdecks to grab his wetsuit, the first waves of nausea washing over him. He would fight the seasickness by jumping into the bracing fifty-five-degree water. But in the cabin, as is the usual case, the feeling only intensified. George zipped up his suit, grabbed his surfboard, and leapt over the gunwale, simultaneously and spectacularly spewing his breakfast into the deep blue sea.

The buoys stayed anchored, offering the surfers a point of reference and a measure of relief. When the next set of four or five waves broke, they showed that the surf was perhaps twelve to fifteen feet from top to bottom. It wasn't gargantuan, and hopefully someone might actually be able to ride one. But that someone would not be George. He lay on his back, prostrate on his board, and staring up at the sky semidelirious, while the California current carried him south at one and a half knots. Sharp eyed his fellow editor with at least a small measure of concern, but he knew that George had been in worse positions, and besides Flame had taken a position on deck with his camera. He'd at least glance at George occasionally.

"He'll be back in a minute," Hulse said. "Let's get out there."

Sharp and Hulse leapt over the side and immediately started paddling across the two-hundred-yard gap between the boat and the wave. The freezing water seeped through the seams in their wetsuits, inducing an involuntary shudder, and the sounds of boat and buoy quickly faded into a strange, muffled silence so complete they seemed to have entered a cave. That is, until the first wave of the next set blurred the horizon just ahead and its concussion split the air like an artillery shell, vibrating the beads of water on the decks of their boards. This was the strangest paddle they had ever made.

A jumpy Sharp kept telling himself not to turn around. He explains, "Surfers are used to looking out to the endless sea, but when you turn around you expect to see the shore. When it's not there, it's instantly disconcerting. Then you'd start to look down, and you realize you don't want to do that either. The water was this deep cobalt blue. You could see thirty, forty feet down into the kelp, where you knew there were sharks the size of submarines. It was just so surreal."

"You gotta understand something," Sam George says. "There was no shore-break, no white water between sets. Nothing. It was silent and flat as a lake. Then these waves come in. It was like the scene from *Jaws* where the shark would come up and scare Chief Brody and then slide back down in the water."

Sharp and Hulse triangulated using their homemade buoys and took a position just to the east of a spot of water that boiled and surged ominously with the passage of smaller swells. You saw boils like this at most big wave spots. It meant the water was swirling around and through caves, boulders, or some other big obstacle. If you crossed one during a hard turn, your surfboard could slide out from beneath you like a snow ski hitting a patch of ice. A sea lion popped up a few feet outside, taking in one of the more bizarre sights in its open-ocean life and inducing a whimper in an already edgy Hulse. As it dove, a wave lurched in—an azure lump about the size of an Olympic swimming pool. They paddled over it and hooted. Another followed immediately in its wake, and another.

"All the things you're used to doing: taking in a lineup from the beach, measuring how far you've paddled according to the beach, duck diving, sitting on the outside because of a crowd—all the things you measure waves by—not one of those things was there," says Hulse. "And you could not see the approaching waves very well—you had to use the *top* of the first wave just to see the second wave. It just lifted up right in front of you. And everything was in motion—the boat and the buoys—everything. I remember just sitting out there after the set passed and thinking, *We're in another world.*"

As if to punctuate the unreality of the morning, the stillness was suddenly shattered by a deep roar. Sharp first mistook it for an undersea earthquake. "A-10 Warthogs," he says. "Tankbusters. These military jets came roaring in, like twenty feet off the water and tipped their wings and turned past us. We could see the pilots clearly, and I was thinking, *Man, those guys are crazy.* But then, they were probably looking down, too, and saying what the hell are those crazy guys doing *down there?*"

Sitting in the water, Hulse turned to Sharp and said, "I think we're just going to have to see one break, get to that spot, and catch the next one. Just go for it, and see what happens."

As if on cue, the horizon darkened again. Hulse paddled over the first wave, using the point where it had crested and a particularly big boil to line up for the second. Then, raw instinct took over. He grabbed the outer edges of his 8-foot 3-inch board, sunk the tail vertically, and then, using the boost of his board's buoyancy, scissor-kicked and whipped around 180 degrees to launch himself in the direction of an imaginary shoreline while immediately windmilling his arms. To lasso a swell moving at twenty-five to thirty knots, you need *at least* five knots of self-generated velocity—preferably more. The wave overtook Hulse a few short seconds later and angled him straight down a rapidly steepening foothill. Acceleration was instantaneous, and the smooth fiberglass base of his board rose to a plane. With two final explosive strokes to seal the deal, Hulse leapt straight to his feet, immediately placing most of his weight on his back leg. This prevented the board from nose-diving and allowed for a quick, sharp turn to his right. He angled hard off the bottom of the wave, unwittingly allowing his right hand to skim along the mirror surface. He rocketed along, staying just ahead of a maelstrom of white water gnashing at his heels. George Hulse was surfing at the Cortes Bank.

"It wasn't a heavy, adrenaline wave," Hulse recalls. "But there was definitely this feeling of incredible speed—of how quickly you were moving down the Bank—like moving down a conveyor belt. I guess because the waves were coming out of the open ocean."

In fact, the waves were moving around 50 percent faster than even comparable waves at Todos Santos or spots along Oahu's infamous North Shore.

Hulse carved and swooped and S-turned for a couple hundred yards. After passing the boat, he kicked out, amazed at how far the wave had carried him along Bishop Rock's shallow perimeter. Sharp scratched into the very next wave and rode nearly as far.

Hulse paddled over to Flame, not sure what to make of the ride. The wave had been astonishingly fast—faster than anything he'd ever ridden at a comparable size. Hulse only wished it had been steeper and more critical, which would have given the world's most demanding surf photographer a more radical shot. But Flame looked as happy as a clam. "You got it," he said, offering a big high five before Hulse paddled back out to the lineup.

Triumph was soon overshadowed by alarm. A set of waves marched onto the reef far outside and bore down. They were impossible to catch and would be impossible for the surfers to avoid. Flame's captain gunned the boat's engine and ran for deeper water just off to the west of the peak, while instinct again took over for Sharp and Hulse. Being caught inside involved the same drill whether you were a hundred miles out to sea or at Waimea Bay. They took three or four short, shallow breaths to fill their bloodstreams with oxygen, cast their boards to the side, eyeballed the craggy bottom and dove deep, saying a little prayer that the thin urethane leashes that bound ankle to surfboard would hold.

The first drubbing was lengthy but not as severe as they feared, a fact Sharp attributed to the deep water beneath the waves. After about twenty seconds in a violent spin cycle, each surfer corked to the surface with lifelines still attached and eyes wide open. Yet when the next wave came and the cycle repeated, Sharp had a panicked recollection. The chart guide had indicated a shipwreck *right here.* Maybe he was somersaulting right above it. He and Hulse were tumbled and spun down the reef, a good hundred yards farther inside from where they started. Another came. Eyes open, Sharp dove for the black bottom—he decided it was better to find what was down there on his own than to meet it involuntarily. With the churning foam, though, he couldn't see a damn thing. When the fourth wave had at last spent its energy, he and Hulse sputtered to the surface, reeled in their boards, and paddled back to the lineup, quaking with adrenaline.

Three or four more midsize sets offered up a few more rides in the ensuing twenty or so minutes, and then the conveyor belt simply, inexplicably shut down. The most likely explanation was that the tide had risen too high for the swell to break.

Sharp and Hulse returned to the boat in silence while the truth sunk in. They had surfed the Cortes Bank *on the smallest wave it was capable of producing.* If a swell was any smaller, it would simply roll over the Bishop Rock without breaking. A swell even five feet bigger, with ten or fifteen waves per set, would present a frightful, perhaps unconquerable challenge—at least given the current state of technology. Not only would the swirling water make it incredibly difficult to position yourself to catch a wave, but the biggest waves would break so far out that the surfers would face deadly walls of smothering white water and a trip to the bottom. Cortes Bank wasn't just a secret big wave spot. It was a big wave spot that only broke at a minimum of 15 feet. The surfers were left to speculate about the maximum wave height Cortes Bank could generate. If the photos Flame took in January were any indication, this might be the biggest wave on Earth.

"You know, even at that relatively small size, it was beyond any scale of any surf spot I have ever seen—like something out of *Waterworld.*" Sharp says. "It was obvious to me that paddling into a really big wave out there was going to be incredibly difficult. But God, the potential. If it had been even 40 percent bigger, we would have gotten our clocks cleaned. There was a kind of recognition that if you went and tried to paddle out on a big day, you would die for sure."

George and Sharp were itching to tell *their* readers the story of their first sighting of surfing's great white whale. But when the Black Watch reached Newport Harbor early the next morning, Flame faced everyone and said, "Look, I want this mission kept secret." He was already planning a return with a crew of A-list big wave surfers in a bigger swell—little did he know that that mission would not happen for better than a decade.

"You can just imagine the angst," says Sharp. "Sam and I basing our entire lives around sharing these experiences with the entire world. To have gone out and done this landmark thing—but we can't tell anyone."

George laughs. "Bill and I have the two of the loudest voices in surf history, and we said nothing."

Chapter 2:

ONCE

UPON

AN ISLAND

"Queequeg was a native of Rokovoko, an island
far away to the West and South. It is not down in
any map; true places never are."

—Ishmael, from Herman Melville's *Moby-Dick,* 1851

Is it possible that what drives big wave surfers to hunt their quarry at Cortes
Bank—a dangerous, monstrous confluence in the open ocean beyond human
reference or scale—resembles the impulse that drove the first people to ever visit
it? We can't know. Evidence of the very first people to settle along the Southern
California coast ten thousand years ago is modest, and written accounts of their
culture are virtually nonexistent. It may be that no ancestor of California's indigenous peoples ever set foot on Cortes Bank, but it would have been possible. And
if big wave surfers are any guide, if it was possible, no matter how difficult, it's
likely that someone tried. Until it was steadily submerged by the slowly melting
glaciers of the last ice age, Cortes Bank was an island, and we know just enough
about the region's original inhabitants that we can speculate what a voyage to it
may have been like.

In constructing this imagined journey to the ancient Cortes Island, I am
deeply indebted to the following researchers: oceanographers Gary Greene and
Rikk Kvitek; geologists Judith and Paul Porcasi; archaeologists Roy Salis, Ellis T.
Hardy, Collin O'Neill, Andrew Yatsko, Clement Meighan, Michele D. Titus, and
Philip L. Walker.

Around ten thousand years ago, seafaring peoples established a permanent
community on what is today known as San Clemente Island, around sixty miles
west of the mainland town of San Clemente. These original inhabitants probably called their island home Kinkingna or Kinkipar, and they were the ancestors
of contemporary Tongva and Chumash Indians in Southern California. Their
home was an arid, twenty-one-mile-long volcanic uplift whose dinosaurian

spine remains easily visible from the mainland and winks like a mysterious beacon for wave-hungry surfers along the crowded California shoreline.

Today, San Clemente Island holds roughly seven thousand documented archaeological sites—a density greater than any comparably sized spot in North America—and they provide evidence of a Kinkipar society of apparently prosperous abundance. The Kinkipar subsisted on cactus fruit, acorns, pine nuts, wild cherries, gritty island tubers, and a turkey-size flightless duck that once swam between all the Channel Islands. But mostly, they were expert fishermen. They dove for white, pink, and red abalone and lobster and were highly skilled anglers who invented every manner of snare, trap, and line-based tackle—catching sheepshead (their primary finned staple), albacore, yellowtail, and shark. The swordfish was the most highly revered for its immense power and magic, and Kinkipar have been found buried alongside the very swords they earned in battles with the mighty creatures.

North of their island home, Kinkipar could hunt pygmy mammoths by paddling to the ancient island of Santarosae: This is what archaeologists call the single landmass that once connected Anacapa, Santa Cruz, and Santa Rosa Islands (southwest of Santa Barbara). Apparently, at some ancient juncture, a small posse of hungry wooly mammoths decided to snorkel five long miles from the mainland to Santarosae, where they established a colony. The pachyderms soon gobbled up most of the food, and scarce resources shrunk their progeny in size until they became a hardy subspecies. About six thousand years ago, the last of these tiny elephants were hunted to extinction.

The same fate eventually awaited the Kinkipar themselves, of course, once North America was discovered by Europeans, who brought disease, acquisitiveness, and war in their wake. In 1542, Juan Rodriguez Cabrillo became the first European known to have explored the coast of the Californias, and in his log he records meeting the locals on either San Clemente or Santa Catalina Island:

"They went with the boat on shore to see if there were people there; and as the boat came near, there issued a great quantity of Indians from among the bushes and grass, yelling and dancing and making signs that they should come ashore; and they saw that the women were ninning away; and from the boats they made signs that they should have no fear; and immediately they assumed confidence and laid on the ground their bows and arrows; and they launched a good canoe in the water, which held eight or ten Indians, and they came to the ships. They gave them beads and little presents, with which they were delighted."

Through the rest of the sixteenth century, Spaniards plied the West Coast in galleons. In *The History of California and the Southwest*, Fray de Zarate Salmeron refers to the arrival of Sebastián Vizcaíno in 1602 to Avalon Harbor on Catalina Island, which the locals called Pimugna.

"The inhabitants of the island made great rejoicings over the arrival of the Spaniards. They are fishermen, using boats of boards; the prows and poops high and the middle very low. Some will hold more than twenty persons. There are many sea lions, which the Indians hunt for food; and with the tanned skins they all cover themselves, men and women, and it is their usual protection. The women are very handsome and decent. The children are white and ruddy, and very smiling. Of these Indians, many wished to come with the Spaniards; they are so loving as all this. From here follows a line of islands, straight and orderly, at four to six leagues from one to another. . . . All have communication with one another, and also with the mainland."

In 1910, author Charles Frederick Holder described a poignant archaeological dig in *The Channel Islands of California: A Book for the Angler, Sportsman and Tourist*:

"At San Clemente, one find I saw Mexican Joe carefully cut out of the damp sand from near a man's skeleton, was a flute, made from the leg bone of a deer. The native had covered it with bits of beautiful pearl (abalone), fastening each piece on by asphaltum, the result being a rude mosaic. It was difficult to consider this aesthetic musician—whom we dug out carefully and sent to the Smithsonian—as very much of a savage. He was buried in the sand dune in a sitting position, his arms bound to his knees, on which rested his head, while in front, behind, on each side, and over him were flutes, each carefully placed, and bearing the beautiful abalone mosaic. Here rested some savage Mendelssohn of the Isles of Summer."

However, by the early to mid-1800s, ruthless Alaskan fur traders had spread to most of the Channel Islands, decimating humans and wildlife with guns and disease. The work they left undone was finished by Catholic missionaries in what the California Indian nations call the Spanish Mission Holocaust. So complete was the erasure of Kinkipar's ancient culture that no living record of a language, religion, or custom survived. Today, San Clemente Island is under protective stewardship of the U.S. Navy, and it's largely off-limits to everyone but military personnel and scientists, who have been given vast, undisturbed tracts to study these ancient Californians.

From the top of San Clemente, on what we today call Mount Thirst, the early Kinkipar would have had a dazzling view of their world in all directions. And it could hardly have escaped their notice that a pair of small, low islands were visible due west at the horizon's edge. The nearest of these would have appeared quite small, the farthest a bit more substantial. Neither was so far away as to be unreachable. The Kinkipar regularly paddled the thirty-two miles between Kinkipar and Harasa (Santa Catalina Island), and Kinkipar navigators would have reckoned that the smallest island was about that far out. The most distant island was perhaps forty miles out.

Any decision to set out for the islands would not have been made lightly. What, indeed, would have been gained by risking death and the loss of precious boats on such a journey? The Kinkipar had plenty to eat on their island and in the surrounding waters; besides, hunger would have almost certainly driven them to the mainland, not further out to sea. There were surely a bounty of otters and sea lions near their home, so perhaps the promise of more fur trading with their neighbors offered some motivation. Perhaps they hoped to find new islands to claim, on which to expand their society, but these islands would have been too far away for regular two-way passage, and they shimmered where the seas were most treacherous and violent. Pragmatic rationales would have likely paled before the known dangers, thus making the most compelling reasons almost assuredly emotional, or perhaps spiritual: The islands existed, and so it would have been impossible not to visit them, whether for pure adventure, to test one's mettle, or simply to put a reassuring label on the unknown. They went because they were there.

At least, the Kinkipar already possessed the sacred craft to make such a voyage. They paddled high-sided boats called Tumol or Ti'at. These vessels linked island and mainland and were no less important to early Californians than the koa wood outrigger canoe was to ancestral Hawaiians. Ti'at ranged from eight to thirty feet in length and featured a steeply raked bowline that helped negotiate swells. They were built of planks of pine or highly revered logs of the mysterious, giant redwood, which washed down the coast from great rivers to the north. Wood was meticulously sawed, carved, and shaped into fitted sections with blades of obsidian, quartz, and bone, then sanded with sharkskin. The planks were then pieced together through drilled-out holes and sewn together with perhaps a mile of cordage wound from milkweed, yucca, and animal sinew. The wood was then sealed and cemented with yop, a pungent mixture of pine pitch and asphalt that washed onto beaches or was dug from natural deposits like those found along the shores of modern-day Carpenteria or the La Brea tar pits. The boats were stained with red ochre, inlaid with abalone and other jewels, blessed by a shaman, and put to sea.

A big Ti'at was propelled by as many as eight men who bore long, double-sided wood paddles. It could carry perhaps four thousand pounds of cargo. Considering the series of long, perfect point breaks and mellow longboard waves that stretch from Point Conception south to Trestles, it's difficult to imagine that the best paddlers weren't adept at surfriding—either for pleasure or survival. When the seasonal swells grew big, they still had to reach shore.

Choosing the correct time to go to Cortes Island would have been essential. A scouting mission would be too dangerous in the spring and summer, when raw northwesterly winds and steep, unpredictable chop were a constant threat. During autumn and early winter, the northwesterlies typically wound down, but smoky easterlies often blew in unpredictably from the mainland, sometimes ratcheting up to gale force. The least threatening window opened during the early winter. This was the time of year the gods hurled huge swells in from the north and west, but once you cleared the nearshore waves, they were easily navigated—provided the weather was fairly benign, and you knew what shoals and reefs lurked on the bottom.

As soon as the Kinkipar chief made a decision to allow a team of his men out to explore the outer islands, he would have asked the shaman to confer with the spirit and animal world. A tea of *tolache*, made from the flowers of jimsonweed, brought visions of hardship, success, or failure. The powerful deity Chungichnish (a known name of a shamanic Chumash God) was consulted, and a tribal dance held to curry his blessing. With clairvoyance granted by the *tolache*, the shaman asked the souls of islanders who disappeared at sea to weigh in on the perils ahead. The shaman ordered the party to carry one of the tribe's wise, surly old ravens. Raven saw the whole world and would alert the mariners to dangers over the horizon. Porpoises were guardians of the world below. It was prudent to ask for their blessing as well. Prayers and offerings begged the winds to lie still and for the sun to shine.

In practical terms, it was a full day's hard paddle to the first island. The journey called for two midsize boats, each manned with four of the island's most able watermen—perhaps younger members of the canoe guild, fishermen, and an elder navigator. They would need baskets of fresh water (yop made an excellent, if pungent, waterproof lining), cordage, spare planks of redwood, and a load of dried abalone and vegetable-based provisions to augment fish and crustaceans they'd hook and spear. Since the Kinkipar never saw boats rowing to or from these islands, the chances of meeting people were slim to none. Still, should they encounter angry elephant seals or demons, prized swordfish spears were also loaded.

The islands were low-lying, so locating them from sea level would present some challenge, but Kinkipar understood celestial navigation and wayfinding across trackless water. Their route would be relatively easy because they could triangulate the first island against Sky Coyote and the Guardian (the North Star and Ursa Major). But they would also rely on an innate knowledge of the known velocity of the northerly flowing California countercurrent and the flight patterns of seabirds that followed straight-line courses toward their land-based roosts.

The scouting party departed Kinkipar at sunset on a calm evening. The men were all relentlessly strong paddlers thanks to endless exercise and a calcium-, protein-, and fluoride-rich diet of seafood. Lengthy exposure to cold water, ceaseless northerly winds, and California sunshine were manifesting in a pair of conditions West Coast surfers would immediately recognize: Bony exostoses were growing to close and protect ear canals, while fleshy pterygia spread from the inner edges of the eye toward the pupils. If these men lived long enough, they'd eventually be rendered deaf and blind.

The crews alternated sleeping and eating, while maintaining a steady two- to three-knot pace. By the time the moonless quilt of stars gave way to dawn, a silhouette of low ridgeline loomed in the distance, and the raven was released from his cage as a scout. Dolphins and sea lions regarded the men with fearless curiosity. The Ti'at finally squeaked onto the white sand of a small cove along the eastern shore of the smaller, closest island—today the submerged feature we call the Tanner Banks.

A hike to the top of a 75-foot-high ledge of hard, black rock revealed an atoll of strange and utter beauty. A long, low, and narrow ridgeline stretched to the northwest and southeast for a couple of miles, eventually bending into a broad oval. In the center lay a vast lagoon ringed with white sand and filled with thousands of squawking seabirds. Along its near shore, a small troupe of otters basked in the sun, bashing open clams with rocks.

A pair of narrow openings a mile distant allowed in a small surge of ocean and swell. At their edges, hundreds of elephant seals lounged and bickered on a narrow sandy beach. The highest of these rock hills rose at the south end of the lagoon, standing sentinel over a smaller lagoon perhaps five hundred yards across. At this lagoon's curving entrance unwound a flaw-less, chest-high right-breaking wave with a thin, translucent lip. The wave was almost identical to the one that reeled across the cobblestone beach that mainland tribes then called Humaliwo, and which later Spanish destroyers would one day call Malibu.

The most telling aspect of this peculiar island, however, was that it was almost entirely devoid of vegetation and surely offered no fresh water. The west-ern flank had been beaten considerably lower than the eastern, clearly the work of ferocious, scouring winds and even more destructive, gigantic surf. The west-ern shoreline of Kinkipar did regular battle with furious waves, particularly in the wintertime. How high would those same swells rise out here?

The party hiked south to the highest overlook, a broad-based, rocky sum-mit of roughly 150 feet. The land here was more substantial and in the pock-ets of sandy soil, familiar dry grasses and sage scrub clung precariously to life. The smaller lagoon below also held water of fantastic clarity. In the tidepools at its edge crawled a riot of sea stars, anemones, limpets, mussels, and enormous white abalone twenty or more years old.

A walk along the western ridgeline gave the men a better look at the entrance to the lagoon—to determine if a Ti'at might navigate inside. As they trekked above a narrow sandy beach, the men were amazed. No animal, not even the ducks, displayed an ounce of fear.

The lagoon's waters were considerably warmer than the ocean. They swarmed with toothy sheepshead, seabass, and bright gold garibaldi, while huge halibut, stingray, and leatherback turtles stirred the sandy floor. Tiny shrimp rocketed away from disturbed stands of eelgrass and brown pelicans dive-bombed into a shimmering swarm of baitfish.

The southerly opening to the lagoon was the narrowest—fifty or so steps across, but aside from a rock outcropping in its center, it appeared free of obstacles. The opening, a five-minute walk to the north, however, appeared considerably wider and marginally deeper—an easy entry point for a Ti'at.

A glimmer of spray turned heads offshore. A trio of orca rocketed into the mouth of the lagoon. Ahead of them, a panicked great white shark. An angling orca rammed the shark broadside. Geysers of blood roiled the water's surface. Should any men ever paddle into this lagoon, it would be best to stay in the boat.

It was agreed; this island was not terribly far from home. Were passages made during fair fall or winter weather, it presented an opportunity for hunting and fishing the likes of which simply did not exist on Kinkipar. The Gods had created an oceanic eden.

The team easily spotted the craggy rise of the next island from the hilltop above the lagoon. Its summit lay tantalizingly close, a paddle of five, maybe six hours. No need to leave just yet. A bundle of wood was fetched from a boat and a ball of yop helped bring it to a blaze—the first fire this island had ever seen. Sunset prayers and an offering of the largest abalone were made to Chungichnish. At nightfall, a star bearing was established, but in the rising moonlight it would be largely unnecessary. The men would sleep for several hours and then set off. Their fire was easily seen from the hills of Kinkipar. It was a good omen.

An hour after midnight, the Ti'at were slicing toward the unknown shore. Paddles trailed eerie swirls of phosphorescence, and the men silently pondered legends of sea demons and giant squid that could pull a canoe to the depths with the sweep of a tentacle. Just ahead a ghostly green tornado of schooling sardine appeared. Moments later, a giant shape exploded from the glow, casting a hissing wake over the bows of the Ti'at. A feeding humpback whale had nearly sunk them all. Perhaps the very God who created these tribes lived on the approaching island. Perhaps this was a warning.

There were subtle changes in air and water, marking new currents. A whisper of breeze put the slightest ripple on the surface, the light slaps pattered against the boats like rain. The ocean cooled suddenly and noticeably as they left the southerly countercurrent and crossed into the vast, swirling river that swept from north to south. They were now paddling downstream, at a combined pace of nearly five knots. A thin haze formed at the ocean's surface as the shadowy peak of the next island began to obscure the stars on the horizon.

An hour later, a cold fog folded over the Ti'at like a burial shroud. Still, the elder navigator had a solid bead on the hilltop—yet a few miles distant. Simple, straight paddling would bring the party within earshot of breaking surf in an hour at sunrise. Yet five minutes on, all heard the crack of a breaking wave—a wave that shouldn't exist in the open ocean. The air carried the dank scent of rotting kelp, and the swells seemed to suddenly come from several directions. They couldn't see below, but clearly the water beneath them was unexpectedly shallow. In a bigger swell, perhaps dangerously so. They were above another island, this one already beaten just beneath the sea by the waves. To their west, the ocean dropped off to unknowable depths. Any swell of substantial size would career along this depthless seafloor with terrifying force. They were running over what we now know as the Cortes Bank's northern plateau.

The navigator ordered the boats to swing around toward the east and follow what seemed to be the shoal's deeper, kelp-free perimeter. The occasional cracks faded over their shoulders at roughly the same time another sound materialized in the fog—the low, steady rumble of surf breaking on a beach. The first hints of gray daylight offered little comfort, and the men dug with grim purpose.

The two boats passed a few strands of bull kelp and again found shoal water. Current led them along an undersea ridgeline that ran toward the sound of breaking waves. It was in the deeper expanse to the east that the Ti'at crossed a series of eddies that marked a sudden return to the warmer southerly countercurrent and the abrupt edge of the curtain of fog. A dark summit loomed dead ahead. They freed the raven and paddled beneath the rising sun.

The men beheld yet another alien world—what is now known as Bishop Rock lay directly ahead. The outline immediately reminded the elder navigator of the shaggy mammoths that roamed the island hills to the north. The three-hundred-foot-high peak possessed nearly sheer sides and a dark, domelike summit whose updrafts were ridden by circling cormorants. From a half mile's distance, it became evident that the dome was not actually smooth, but a tortured, gothic landscape of vertical spires and canyons. The peak held scores of moving white dots, which on closer examination proved to be the heads of nesting bald eagles. At its base, a series of caves burped, steamed, and burbled in the swell.

Another half hour's paddling gave a view around the island's eastern flank. The black dome transitioned more smoothly down to a flatter plateau whose shoulders then slumped toward the ocean. The island swept southward, perhaps a half day's walk in length, with another dark mesa looming halfway down. A broad, sandy beach ran along most of the shoreline below low cliffs of blond sandstone. But there would be no landing here. The beach was blanketed with miles of angry, bloody male elephant seals, each battling for a small roosting spot. The waters here swarmed with shark, barking fur seals, and dive-bombing eagles and pelicans. The men paddled back around the dome and through the shallows along the island's north end. Perhaps they could find a protected landing along the western flank.

Yet the entire western shoreline was being pummeled by closely spaced, rugged breaking waves. The swell was not very big, only a few feet, but it would be impossible to tell how high the breaking waves actually were until you were in front of them. And once there, you were committed to a shoreward course. It was duly noted that the lowest stretches of land were covered with beautiful dunes that bore little vegetation. This meant regular scouring winds and perhaps an ocean that occasionally washed clear over the lowest stretches of beach. The scouts could probably bring their Ti'at in through the surf just ahead, as there were no elephant seals, but it seemed too risky. With only two boats, losing one in the surf would be a disaster.

It was by now midday, and they paddled far enough offshore to partake of a meal. They discussed their situation and what they'd seen, bobbing together in the waves. This island clearly held the promise of clothing and trade in otter and fur seal, protein in the form of plentiful swimming ducks. The confluence of warm and cold waters between the eastern and western shoreline surely promised epic battles with swordfish and perhaps bluefin tuna. It was agreed they would return to these islands, but with more boats and men. They should land on the more protected leeward shore when the elephants had returned to the ocean. For now, it was best to paddle back to the protected side and its warmer waters for the paddle home before the fog rolled back in. It had been burned back by the sun but hovered in an impenetrable, wraithlike line about a mile offshore.

Rounding the northwestern edge of the black dome of rock, the Ti'at began to rise and fall almost imperceptibly over undulations that took nine long breaths to pass—a new swell. These waves ran fast and deep but didn't break. Strong eddies began to swirl beneath them. This water was too shallow.

The men turned to paddle straight offshore, but they battled thick mats of kelp and a northerly current grown stronger with the rise of swell and drop of tide. The old raven perched on the bow of the lead boat, crowing orders at men who now rowed as if their lives depended on it. A new line of slate gray lumps surged in from the fog. The first steepened sharply, but rolled beneath the canoes without breaking. At its summit, tentacles of kelp released their grip, enabling faster paddling. The second wave was even bigger. The raven took to the sky, calling encouragement.

Both teams sprinted to get over it. The elder pilots of the first canoe were outflanked by the younger scouts of the second. The wave formed a terrifying ridgeline, yet both Ti'at had a good head of steam. They powered up a near-vertical face as tall as three boats stood end to end. A gale of mist blew off the wave's back, drenching the men and producing the curious sound of a rainstorm on a still pond. White water erupted in their wake. A third wave came, more massive still. The young scouts paddled with all their might as the wave crested, and they uttered great war whoops as their bow crashed down safely on the backside. Breathless with fear, they turned to urge on their elders. But the other boat was gone. Three more waves came, each bigger than the last.

When the waves subsided, the terrified scouts made a fast, desperate paddle back toward the boiling impact zone. The elders were tough, long-winded watermen who knew how to survive. You couldn't spearfish, abalone dive, or canoe off Kinkipar without occasionally dealing with terrible seas, and each had thus faced similar, if not quite such massive conditions many times. You breathed short and fast to flood your blood with air, dove deep, said a prayer, and gave yourself up to the wave. If the Great Spirit was willing, you would eventually be roiled to the surface. But the scouts found nothing. No men. No boat. No supplies. The God at the edge of the world played for keeps.

The fog folded over the remaining canoe like a death shroud, and the scouts dug back out to deep water as a new set of even bigger waves thundered down behind them. The raven turned an arc in the clear air above the mist and flew back to the black rock. He planned to circle around in the updrafts with the cormorants and eagles, gaining as much elevation as he could, and then aim for Kinkipar, whose summit he could plainly see in the distance.

The scouting team had ventured out to the edge of the world, to a sacred realm where the mortal dared not tread. They had, it seemed, reached the defined limits of human exploration. If the younger scouts were very lucky and made it home, their tale of the strange islands and their perilous journey might be told around Kinkipar firesides for thousands of years. If they never reached Kinkipar, the raven would help the shaman divine the details. Amid the wailing and heartbreak, heated debate would surely ensue. The cautious would argue that the islands were the sole domain of the Gods and the chief should pronounce them forever off-limits. The daring would make the case that the islands promised adventure and riches, and besides, what if someone had survived and was still out there? Regardless, the deadly islands surely continued to call to the Kinkipar like sirens even as they slowly slid into a rising sea.

Eventually, the islands were buried by the waves and faded into Kinkipar legend. Yet humans did eventually return, using bigger, stronger boats that made it possible to approach a mysterious lost shoreline and to taunt a giant rising from the depths. Like the Kinkipar, they simply couldn't help themselves from wanting a closer look.

Chapter 3

PAWNS

TO BISHOP

ROCK

"I must shun this island of the Sun, the world's delight.
Nothing but fatal trouble shall we find here."

—Odysseus, from Homer's *The Odyssey*

In the eons after San Clemente Island was first settled, massive melting of the Greenland and Laurentide ice sheets poured trillions of gallons of water into the oceans. It became impossible for Kinkipar to land atop Tanner Island perhaps eight thousand years ago. When the Great Pyramids were completed, around forty-five hundred years ago, all that remained of Cortes Island was a wave-scoured dome perhaps ten feet high and a half-mile wide. By the mid-nineteenth century, Kinkipar culture had essentially disappeared. So had Cortes Island.

Through these millennia, this uplifted mesa a hundred miles from the mainland transformed from a tantalizing destination for possible human settlement into a fisherman's eden and a perilous shipping hazard, one that produced a truly frightening wave. Just how many ships might have been destroyed, before the mid-1800s, by some combination of enormous waves and the Bank's shallow seafloor is, and probably will always remain, unknown. The first mapping of the California coast began with the seminal trek of Juan Rodriguez Cabrillo in 1542. But no Spanish map seems to have recorded the existence of either Cortes or Tanner Banks. In the 1981 printing of *Shipwrecks of the Pacific Coast*, James A. Gibbs noted that the gold-laden galleon *Santa Rosa* met her fate in 1717 atop the Bishop Rock in a nightmarish specter of foam and splintering wood. The book's first 1957 edition makes no mention of the wreck, so how Gibbs later received this information, and its veracity, is one of the Bank's innumerable mysteries.

Indeed, even when it comes to the verifiable, written documentation that has been uncovered to date, the early modern history of Cortes Bank is rife with falsehoods, legends, and oversize characters. Its mystery and menace are writ larger than life—even its charting and naming. Once its territory *was* properly staked and mapped, by a charismatic but doomed man, its miles of sandstone and basalt would serve as a dangerous playing field for daring and ambitious mariners of all stripes.

The first person we *know* to have recognized something unusual a hundred miles out was an American crewman aboard one of the most famous ships ever put to sea. James Alden was born in 1810 in Portland, Maine. He was a direct ancestor of *Mayflower* pilgrims John Alden and Patricia Mullins, a young woman reputedly cast into a love triangle between Alden and the *Mayflower's* captain, Miles Standish. (The affair was immortalized in Longfellow's scandalous nineteenth-century poem *The Courtship of Miles Standish*.)

Alden was ballsy, irascible, and a talented navigator. During his early years, he was assigned to an epic survey of the South Seas under Commander Charles Wilkes. The mission charted some fifteen hundred miles of Antarctic coastline and made expansive forays through the South Pacific. At the expedition's end, Alden was assigned as an officer aboard the USS *Constitution*. This legendary, seemingly indestructible frigate earned the name "Old Ironsides" in the War of 1812 for the confounding ability of her oaken hull to deflect British cannonballs. On the *Constitution*, Alden served alongside a famed and unbalanced captain, Jonathan "Mad Jack" Percival. Percival was a sailor's captain, at least some of the time. During an 1826 visit to Hawaii, he risked his career by demanding island chiefs rescind a missionary-sponsored law that forbade local women from boarding ships, so as to ease the sexual tensions of visiting sailors, and yet later he became known for taking an almost maniacal glee in ordering subordinates flogged or even pistol-whipped at the slightest transgression.

Alden and Percival sailed the world aboard the *Constitution*, becoming, among other things, the first Americans to attack Vietnam. Then in late 1845, rumors of war between the United States and Mexico reached the *Constitution* in Hawaii. Percival led a miserable passage that reached Monterey, California, on New Year's Eve. The harbor was bereft of American ships, so the *Constitution* immediately made a fourteen-day run for Mazatlán.

Strong, frigid northwesterlies propelled the ship at a clip perhaps as high as fourteen knots (16 mph). Her logs while off California carry characteristically simple entries that belie little of the drama or misery: "Got up the larboard chain and bent it — got the larboard anchor off the bows . . . pumped ship out . . . inspected the crew at quarters, exercised the 1st Division of great guns . . . Punished Wm Brackley (O.S.) with 12 lashes of the Colt for selling his clothes."

We're thus left to ponder a late-afternoon conversation on January 5 among Percival, Alden, and Lieutenant G. W. Grant as the *Constitution* proceeded along a course that supposedly kept her well clear of navigational hazards. A log entry from Grant offers a tantalizing shard of information: "From 4 to 6 [4–6 P.M.] moderate breezes and clear pleasant weather. At 4-20 [4:20 P.M.] discovered breakers bearing N.E. about 10 miles distant."

The night before this strange sighting, the moon was at half phase. It disappeared below the evening horizon around midnight, leaving the *Constitution* sailing toward an uncharted seamount under a black sky. Were she gliding over a large groundswell, an adept watchman might recognize phosphorescent foam from breaking waves off her bow well before they created a hazard. But waves were not seen until 4 P.M. the next day. This speaks to a likely, sudden arrival of a powerful winter swell.

Percival, Alden, and Grant surely extended telescopes, marveling and puzzling as giant, ghostly white horses galloped across an utterly empty ocean. Wonder would have been mingled with dread and relief. The *Constitution* drew 22 feet of water. Some sunken menace was clearly spawning the massive breakers, yet when the ship had first approached it, it seems nothing had been seen. Had Percival proceeded a little farther west at flank speed, the Cortes Bank would have done the job British cannonballs could not.

After surviving alternating rounds of terrifying combat and ceaseless boredom during the Mexican-American War, Alden was offered a position with the western command of the U.S. Coast Survey. At the helm of a navy steamer, *Active*, he would lead the effort to map the vast swath of new United States territory along the Pacific Coast.

It would be difficult to overstate the danger, importance, or maddening tedium of this work. The discovery of gold in 1849 created an explosion of demand for transit to California. Yet shoals, currents, and tides were largely a mystery, and maps—many of which dated to the seventeenth century and even before—were so dangerously inaccurate as to be worse than useless. Coast Survey captains navigated creaky, leaky ships along a strange and spectacular coast peopled by suspicious and occasionally hostile natives. To compound the thankless duty, sailors regularly abandoned ship for the gold mines. In 1849, gold-hungry crewmen beat senseless Captain William McArthur of the survey cutter *Ewing* and threw him into San Francisco Bay. He was saved only when a fellow officer spotted him drifting toward the Pacific and latched onto his hair.

For anyone with a passing interest in California history, the writings of Alden and MacArthur are fascinating, for therein lie the first descriptions of a once nearly empty coastline today inhabited by millions. For our purposes, the most relevant passage is one Alden wrote describing the Southern California coast:

From the last-named point [Point Conception] to Santa Barbara the coast is almost straight and runs nearly east and west. The passage formed by the islands lying abreast of this portion of the coast is called Santa Barbara Channel. . . .

We steered down to the southward and westwards, so as to get on the parallel of latitude well to the westwards of certain dangers said to exist somewhere between that point and the Coronados Islands. Our search was not entirely unsuccessful, for we fell in with a bank, where the shoalest [shallowest] water we found, however was forty-two fathoms [152 feet], fine white sand. For many reasons, our examination was limited. . . . I should not, therefore be willing to say that there are no dangers existing in that quarter, particularly as one of the shoals that was discovered by the U.S. Frigate *Constitution*, in January, 1840 [Note: this date proved to be a typographical error and should have read 1845], and I happened to be one of the officers who believed they saw it at the time. I shall improve the first opportunity to give that locality a more thorough examination.

Alden would eventually make good on this promise, yet in the meantime, he wasn't the only one to witness the Bank's strange specters. In their quest to reach the California gold mines, countless 49ers traveled over land, but scores more— perhaps spooked by horror stories of parties withering in hellish desert infernos or feeding on fallen comrades in Sierra snowfields—chose the longer but safer route by sea: sailing the Gulf of Mexico, crossing a lawless Panama, and then being ferried north along a route plied by ships that came to be called Panama Steamers.

A January 1853 advertisement from the *New York Times* noted such an opportunity:

The magnificent new double-engine side-wheel steamship CORTES (1600 tons) THOS B. CROPPER, Commander, will be in readiness at PANAMA to receive the UNCLE SAM's passengers and sail immediately for San Francisco. The accommodations and ventilation of the Cortes are all that can be desired. Her speed (established on the voyage between New York and Panama and while on the Pacific coast) is unequaled.

It seems that on this very January journey Cropper witnessed a baffling spectacle. Passing south of San Clemente Island in a heavy swell, he was astonished to find massive eruptions of white water in the middle of the deep blue sea. The *Cortes* ventured close, and Cropper became certain he was above an undersea volcano. Upon reaching San Francisco, Cropper apparently met with

NEW-YORK AND SAN FRANCISCO STEAM-
SHIP LINE.—THROUGH TO SAN FRANCISCO,
VIA ASPINWALL AND PANAMA, at greatly reduced
rates   No detention on the Isthmus. The new and splen-
did side-wheel steamship UNCLE SAM, (being the second
voyage) 2,000 tons, WM. A MILLS Commander, will leave
New-York for ASPINWALL, (Navy Bay,) on THURS-
DAY, Jan. 20, at 3 P. M. precisely, from pier No. 3 North
River, where passengers can examine her accommodations.
The performance of this steamer on her first voyage, war-
rants the assurance that passengers will reach Aspinwall in
from seven to eight days. The magnificent new double-engine side-wheel steamship CORTES, 1,800 tons. THOS. B.
CROPPER, Commander, will be in readiness at PANAMA
to receive the UNCLE SAM'S passengers, and sail immediate's for SAN FRANCISCO. The accommodations and
ventilation of the Cortes are all that can be desired. Her
speed (established on the voyage between New-York and
Panama, and while on the Pacific coast.) is unequal'ed.
    Every exertion wi'l be made on the part of the line to en-
sure comfort, expedition and sa'ety. Passengers will be
landed on the wharf at Aspinwall free, taking the Panama
Railroad, and are cautioned that Tickets for this Line are
sold only at No. 26 Broadway.   . _ _   _   . .   .

Coast Survey officer W. P. Blake, who recounted Cropper's direct words: "The
waters were in violent commotion and thrown up suddenly into columns, at
regular intervals of four or five minutes."

Alden's *Constitution* sighting had been verified. He dispatched the U.S.
Coast Survey cutter *Ewing*, now captained by Lieutenant T. H. Stevens, to inves-
tigate and chart the seamount. Yet for some reason, Stevens did not reach the
strange shoal until mid-May 1853. By then, the powerful swells of the wintertime
North Pacific were gone, replaced by punishing northerly winds and smaller, but
relentlessly choppy seas. Stevens spent five awful days before reporting to Alden:

Sir: I have the pleasure of reporting my return to this place from the
shoal to the southward of San Clemente and San Nicolas, of which I have
made a thorough examination, having been five days anchored upon it.

The shoal, or bank, is latitude 32 degrees, 39 minutes north longitude, 119 degrees, 19 minutes, 50 seconds west. The island of San Nicolas bears NW by N, distant 46 miles, island of San Clemente bears NE
½N distant 43½ miles. The nature of the bottom is hard, composed of
white sand, broken shells and coral. The least water found—ten fathoms—would be about nine, reduced to low water and the character of
the soundings, as you will find upon reference to the chart which I send
herewith, irregular and abrupt.

The weather, while at anchor upon the shoal, we found different from that which normally prevails upon the coast in the vicinity, bearing a strong resemblance to that upon the banks of Newfoundland.

The current is irregular, frequently setting against the wind and running with a velocity of nearly two knots per hour, producing a heavy sea, and causing the water probably to break in heavy weather, as has been reported.

In his correspondence to Alden, Stevens included the first fairly detailed chart ever drawn of the Bank. It revealed a sort of hilly mesa, roughly twelve miles long and six miles wide, oriented in a northwesterly direction, "in the immediate path of the Panama Steamers."

The report seemed to bear some good news. A minimum depth of nine fathoms, or fifty-four feet, would not puncture a steamer's hull, and though waves might break above a fifty-four-foot-deep shoal during a truly monstrous swell, that would be exceedingly rare. Then again, a strong, southerly current of two knots—a half-knot below the average speed of the Gulf Stream—might also be problematic to ships approaching the Bank. When a northerly swell runs headlong into a southerly current, its waves can steepen and stack up against one another into something resembling a line of harried passengers squeezing onto a narrow airport escalator. This is one of the ways so-called rogue waves are spawned.

But Alden also realized that unless he had once seen a procession of giant ghosts from the deck of the *Constitution*, Stevens had likely missed the mark. While he had surely found a remarkable offshore seamount, there must be an unseen, as yet unfound feature that would throw up towering breaking waves—visible from ten miles distance—in lesser swells.

Alden then christened the shoal "Cortez Bank," based on an incorrect spelling of Cropper's ship. Yet in so doing, he forever obscured the ship and crew that actually discovered the Bank: The USS *Constitution*, "Mad Jack" Percival, and himself.

During the next couple of years, Alden and his men laid the California coast bare. Among their brilliant discoveries were mappings of a tortured seafloor of mile-deep canyons and sunken ridges that linked San Clemente Island with Santa Barbara Island, San Miguel and Santa Rosa with Santa Cruz, and tiny San Nicolas to the long rise today known as the Cortes Bank.

During the course of these ensuing surveys, Alden's suspicions about a far more shallow and dangerous rock were confirmed, apparently twice over, during the first half of 1855. However, like Cortes Bank itself, this feature would be misnamed, due to what was in all likelihood the conflation of maritime myth-making—in which a good story grew even better, and the most outlandish version eventually accrued the patina of truth. The other, far more interesting and tragic tale of discovery, on the other hand, has remained essentially unknown.

In April 1855, a fast little clipper ship lay docked in San Francisco, readying for a trip to New York. The *Stilwell S. Bishop* had first been put to sea in 1851 by a wealthy Philadelphia merchant of the same name. She was low slung and sharp-lined, 140 feet long and 31 feet wide. She drew only fifteen feet of water and weighed a mere 595 tons. In 1853, a passage from Baltimore around Cape Horn to San Diego took her only 112 days—a sailing record that still stands.

Detailed accounts of *Bishop*'s journeys are scant. On June 12, 1854, she reached San Diego bearing supplies for a local army garrison and had dropped anchor in Benicia, a port town in San Francisco Bay, by early July. She reached the East Coast in December, picking up a new captain, William Shankland, and had again reached San Francisco by April of 1855. She then advertised cargo space on her next voyage east in an ad in San Francisco's *Daily Alta* newspaper on May 7:

The Regular packet A 1 clipper ship SS Bishop.
Capt Wm Shankland. Will be dispatched immediately. . . . The "SSB," from her small capacity and sailing qualities offers unusual inducements to parties wishing their goods delivered in the shortest possible time. She has made three trips from this port to New York and has always delivered her cargo in fine order. For freight passage, having fine state room accommodations, apply to: H. K. Cummings & Co. 48 California Street.

Shankland departed for the above passage on June 2. He endured ferocious gales that swept at least one ill-fated crewman into the sea—a perilous journey reported in the *New York Times* in some detail. Yet nowhere do the *Times* or *Daily Alta* mention the incident for which the *Bishop* has been immortalized, her collision with the shallowest spire above the Cortes Bank, presumably on that 1855 voyage. As the story goes, after striking rock in the midst of a storm, and with a hull then scratched, dinged, or otherwise splintered, *Bishop* somehow managed to limp five hundred miles north back to San Francisco.

This oft-repeated tale, though, lacks virtually any corroboration, which alone calls it into question. In the mid-1800s, the United States was a maritime nation. Even the most mundane comings and goings of each and every ship were covered by zealous reporters. Yet the story of this collision appears in no newspaper account of the time, which from a sailor or navigator's viewpoint would have been inexcusable. Thousands of souls were now being ferried past the Cortes Bank each month. Had the *Bishop*'s hull been so much as nicked, it would have been the height of irresponsibility for Captain Shankland to fail to immediately notify James Alden and the Coast Survey. Had Shankland remained silent, the news would have almost certainly been slipped by a spooked *Bishop* crewman or passenger and pursued eagerly by reporters for the *Alta* or *Times*.

The timely reporting of such a notable wreck should have also appeared in the scrupulous reports of James Alden, but it does not. The first mention appears in reports of the Coast Survey in an 1858 correspondence in which Captain George Davidson, a colleague and competitor of Alden's, gives the barest mention of the *Bishop* "striking the rock" three years earlier in 1855.

Contemporary historians have puzzled over this incongruity. It was first raised by the editors of *Mains 'l Haul*, the definitive journal of California nautical lore, in 1968. The question has also vexed Steve Lawson, an Orange County treasure hunter and rabid historian who has assimilated vast records of the movements of nearly every ship that plied the West Coast during the Gold Rush. Lawson agrees with a hypothesis laid out in *Mains 'l Haul*. The *Bishop* didn't strike the Cortes Bank at all, especially given her shallow draft.

"Ships sailing toward Frisco during the Gold Rush would have not sailed close to Southern California for two reasons," Lawson said. "First, the prevailing wind and currents along the coast travel south and sailing ships could find better winds a few hundred miles out to sea. Second, the eight islands of Southern California were to be avoided since they created major navigational hazards."

Lawson thinks that if the *Bishop* ever did strike the Cortes Bank, it occurred after she called on San Diego in 1854. "That *could* account for her running relatively close to shore and hitting the Bank," he said. "Then heading for Benicia and bypassing San Francisco makes sense because she would have been sailing in ballast [using rocks to keep her hull low and her keel down]. She wouldn't have a cargo to unload, and there was a shipyard in Benicia where she could be repaired."

Yet Lawson believes that the lack of any documentation of a collision makes this highly improbable. It seems more likely that her captain, instead, merely stumbled onto the same spellbinding white water spectacle witnessed by the *Constitution* and *Cortes*. Then, in its retellings, the *Bishop*'s story morphed into a direct encounter that subsequently lent the feature its name: the Bishop Rock.

By all rights, though, this rock should be named in honor of Archibald MacRae.

In late 1854, a thirty-four-year-old lieutenant accepted a position under James Alden to take command of the U.S. Coast Survey ship *Ewing*, which in 1853 had made the first survey of Cortes Bank.

Archibald MacRae was born in Wilmington, North Carolina, to an American general named Alexander MacRae. Young Archibald cut his maritime teeth in the Atlantic and Mediterranean serving "Mad Jack" Percival aboard a famed navy frigate called *Cyane*. He was a highly skilled navigator, astronomer, and a witty raconteur who regularly wrote home with hilarious, fascinating tales of his life at sea. Short of a pirate, MacRae is as colorful a character as you'd ever hope to meet on the open sea, and it's hard to imagine why he ever accepted his job working for Alden and the Coast Survey. Not long after embarking on the sailor's life, he penned this self-portrait of himself in a letter to his father in 1841: "If, in about eight or ten days, you should see a small character with long red hair, something between the colors of red and auburn, a little lame, a small red nose and several other peculiarly, peculiar peculiarities, you may without hesitation claim the hopeful boy as your son."

In the year leading up to the Mexican-American War, MacRae was sent to Central America as a spy. Posing as a British naval officer, he ferried top secret correspondence between Washington and California—it's a barely known fact that MacRae was the person who actually snuck orders to California's commanders confirming the rumor that the United States was actually at war. After the war started, the wily Tarheel rejoined the *Cyane* under Captain Samuel Francis DuPont. He trekked to Hawaii to protect the US whaling fleet, helped oversee the sinking of some thirty ships off Mexico, and became the deadly leader of a team of U.S. Marine troops that DuPont called his "Fire Eaters." After the war, MacRae made a peacetime traverse of the Chilean Andes with a naval expedition, trekking over two thousand miles of utter and often violent wilderness.

When he returned to Santiago, MacRae wrote to his brother of plans to ask a girl named "Susan" to marry him, though perhaps he was only caught up in the romance of his own adventure. "From the time of our first arrival . . . everything had passed so rapidly as to appear a dream, and I was at a loss for a while to determine whether I had been in Santiago or not; but when I felt my bleeding heart, and got the scent of French brandy from my pocket handkerchief, I knew that all was reality."

Archibald MacRae departed Valparaiso via steamer on January 15, 1854, traveling alone. As to why he returned without a fiancée, we can only guess.

On January 30, 1855, MacRae reported for duty as captain of the *Ewing*. He didn't write home much during this time, explaining he had no good tales to tell. In addition, the logbooks of the *Ewing* and of Alden's ship, the *Active*, seem to have disappeared into a black hole. Personal pictures of shipboard life are thus scant. In 1855, a young officer named Philip Carrigan Johnson served aboard the *Active* with James Alden, and did time briefly aboard *Ewing* alongside MacRae. His diary is one of few personal windows belowdecks.

Johnson describes a posse of inveterate chess players, bird hunts, near drownings, and a mingled annoyance with and great respect for James Alden. He also noted the deep mutual disdain between Alden and fellow survey captain, George Davidson. Alden's wife—a woman already wrecked by her husband's lonely years at sea—lived aboard the *Active*, and Davidson took a rabid delight in terrifying her with tales of the ocean. "I expect she will explode one of these days on account of Davidson," Johnson wrote.

The accomplished and headstrong MacRae, meanwhile, found the leaky *Ewing* something like a brig. In the year before joining the *Ewing*, he had taken ill, racked with recurring, debilitating rounds of a flu-like illness whose cause was a mystery. In the wake of a bout that laid him low for days, he wrote home to his brother Donald:

> Our duty here is so uninteresting and the whole coast so inhospitable that I can find nothing worth relating and therefore must leave it to your imagination to make out what our course has been. Running lines of soundings on a very exposed coast, looking for rocks and indeed suffering every inconvenience except not catching fish, if that may be considered a convenience, when we are so tired of their sight as to loth [sic] them. That has been our last two months experience.
>
> From here we shall proceed near the mission of San Buenaventura [Ventura, California] and probably in a month from this repair to the coast near Monterey. If I did not suffer a good deal with rheumatism and other ailings, I should speculate on what might occur in the future but as it is I can't say.

In the summer of 1855, possibly in July, Alden ordered MacRae to sail for the Cortes Bank, perhaps in reaction to the stories circulating of the *Bishop*. Yet unlike T. H. Stevens, MacRae actually managed to locate the submerged black dome that is the Bank's shallowest reef, and one imagines that, at least for a moment, the deadly tedium of his work lifted just a bit. Even if MacRae did not witness Alden's white horses, he would appreciate the danger to shipping the rock represented, the importance of his discovery, and the deadly spectral waves such a shallow, shaggy undersea cliff would produce.

News of MacRae's findings crossed Panama and reached the director of the Coast Survey in early October. They were published in the pages of the *New York Times* on November 3.

Dangerous Rock on the Coast of California.
From the Washington Union.

The following is a letter from the Superintendent to the Secretary of the Treasury, communicating the position of a dangerous rock on Cortez Bank, coast of California, determined by Lieut. Commanding ARCHIBALD McRAE, United States Navy, assistant in the coast survey:

COAST SURVEY STATION,
DIXMONT, Me., Oct. 10, 1855.

SIR: I have the honor to report that, under the instructions of Lieut. Commanding JAMES ALDEN, United States Navy, assistant in the coast survey, a dangerous rock on Cortez Bank, off the extreme southern coast of California, was sought for by Lieut. Commanding ARCHIBALD McRAE, United States Navy, assistant coast survey, and determined to be in latitude 32 deg. 29 min. north, and longitude 119 deg. 04½ min. west, (both approximate.) The shoalest water on the rock is reported by Lieut. McRAE to be three and and a half fathoms, subject to a possible tidal reduction of six feet, which might reduce it to *two and a half fathoms, or fifteen feet.*

Lieut. McRAE placed a buoy composed of two casks, with a flag-staff between, upon the shoalest part of the ledge to which this rock belongs, and which he represents as quite extensive. The buoy could be seen in clear weather about three miles.

I would respectfully request that a copy of this letter may be sent to the Light House Board, that their attention may be directed to the placing of a beacon on this ledge.

A. D. BACHE, Superintendent

Meanwhile, on October 13, 1855, MacRae either asked or James Alden ordered him to return to the Cortes Bank for a thorough, weeklong survey. This time, Cortes Bank surely impressed upon MacRae its majestic profile and let him witness at least a portion of the North Pacific power it could unleash. He had uncovered a hazard unlike anything off the coast of California. Soundings revealed a series of stair steps fifty to a hundred feet high that told a story of periods of rapid sea level rise and erosion by forces almost unimaginable in scale. Alden and Davidson would come to postulate a truth that geologists would later confirm: Though no active volcano ever rose out here, the Bank's black stone is not only volcanic in nature but is probably still growing. Yet as California's tectonic plates push the Bank upward, toward the water's surface, the more that rock is pulverized by waves. It is a battle that has raged for hundreds of thousands, perhaps millions of years.

Then, if you were to walk off the final stair step—itself some four hundred feet in depth—you'd plunge into abyssal waters MacRae sounded out at better than six thousand feet deep. As open-ocean swells approached the coast, what would they do when they met this mile-high wall? MacRae was both a man of science and a seasoned sailor who had survived towering seas around Cape Horn and visited Oahu during the wintertime big wave season. Despite how little was known about wave formation in 1855, he almost surely surmised exactly what would happen out here: Swells would lift to heights that defied the imaginations of scientists and seafarers.

In the meantime, MacRae was bearing a hardship that no doubt made this expedition both the height of wonder and the acme of misery. He was unable to recover fully from one of his regular bouts of "rheumatism," and he began to complain first of debilitating headaches, and later a persistent, maddening sensation around his right temple. When pushed for a description, he called it "a vibration in my brain."

In early November, *Ewing* left the Bank and reached Santa Barbara, but MacRae was unable to obtain satisfactory medical attention. He ordered the *Ewing* to San Francisco, anchoring alongside Alden's *Active* off Market Street Wharf on November 12. There, an old Wilmington friend named John Savage greeted him. "He was not at all well and he seemed in low spirits and disgusted with everything out here," Savage wrote.

Active's surgeon, Dr. John M. Browne, must have thought MacRae looked a wreck—a bit jaundiced and with dead corpuscles turning his urine the color of cola. Browne described MacRae's ailment as a "biliary [liver or bile] derangement attended by slight fever of the remittent type, both affording symptoms peculiar to a form of fever prevalent at the Lower Coast [a term for South America]."

In short, MacRae had malaria, which he must have contracted from a mosquito bite a couple of years earlier on his South American adventure. Where avalanches, landslides, blizzards, lightening, bullets, mortars, thieves, and giant waves had failed, tiny parasites were at last succeeding.

Dr. Browne administered opiates and ordered MacRae to rest. Three days later, MacRae announced a near-miraculous recovery. The finally cheerful lieutenant spent Friday and the following Saturday enjoying San Francisco with a small, tight crew of North Carolina adventurers.

Yet late on Saturday after returning to the *Ewing*, the dead cells clogging MacRae's brain brought the buzzing, throbbing sensation back with a vengeance.

"If this doesn't stop," he told James Alden's nephew James Madison Alden, who had been working as his assistant, "I'm going to jump overboard."

MacRae was soon surrounded by Dr. Browne and various crewmen and friends. Browne could find no other symptoms to treat, so he gave him more opiates to ease the pain. Then, his friend Savage wrote, "He spoke of bygone days, of different voyages and the time he passed in Chile."

Eventually, MacRae lay alone on his bunk, attended by Madison Alden, who sat in a chair reading the *Daily Alta*. The air was suddenly shaken by a thunderous blast. A massive .44-caliber Colt Dragoon skittered across the floor and bounced off young Alden's boots. He fled in mortal terror, shouting to the others: "The Lieutenant just tried to shoot me!"

"Horror became depicted on our countenances at the terrible spectacle that greeted us," Browne wrote. "Mr. MacRae was seen reclining upon his berth, his face in a bath of blood, the right portion of his cranium blown away, brains and integuments scattered in all directions upon the deck and bulkheads."

On the cabin's desk lay a note penned in MacRae's flowing handwriting.

San Francisco, Nov. 17th, 1855.

As it may be possible for me to leave this world for another, I wish this disposition to be made of my property.

Viz: My watch to my brother John Colin; my funds in equal parts · to the two children of my brother Alexander and the one of my brother Donald.

My books I will to Phelps [MacRae's assistant during his time in Chile], and my body to be taken outside the heads and be dropped overboards on the ebb tide.

If on settling my accounts there should be money due me, I wish it should be appropriated to the purchase of seal rings to be given to each of my brothers.

My soul I give to God and I hope he will make better use of it than I have.

Signed, Archibald MacRae
Lieut, U.S.N.

In the decades after MacRae's death, waves of immigrants and pioneers continued to knowingly and unknowingly risk their lives at Cortes Bank. Most were simply passing north aboard steamships powered by leaky, explosion-prone boilers. Others, though, sought out Cortes Bank deliberately, attracted by the same marine treasures that centuries before dazzled the ancient Kinkipar. Live coral reefs, jungles of palm and bull kelp, deep crevices, and strange, boulder-filled craters created the perfect habitat for black, white, pink, and red abalone and giant, decades-old lobsters. Plying the waters were gorgeous pelagic fishes—yellowtail, yellowfin, marlin, and massive bluefin—so big, fast, and strong they were nearly impossible to catch by lure or spear.

Still, as the twentieth century dawned, the defining aspect of Cortes Bank wasn't its incredibly fertile waters but the perils it posed for navigation. Such was the perceived danger that in 1911, U.S. Navy Commander William Adger Moffett ignored the fact that the Bank actually lies in waters due east of Mexico. He deployed a new high-tech buoy that featured not only a bellows-driven whistle but an acetylene-powered blinking light. He hoisted a U.S. flag, and in a modest ceremony summarily claimed Bishop Rock and Cortes Bank in the name of the United States. The *New York Times* stated that Moffett had actually laid claim to a small island. They were, of course, several thousand years too late for the statement to be accurate, but have yet to issue a correction.

Because steamers didn't have to cruise far offshore like sailing vessels, Bishop Rock's waves were soon being frequently sighted, and the Cortes Bank began captivating writers for the *Los Angeles Times*. In 1925, Maude Pilkington Lukens joined the Coast Survey ship *Pioneer*, whose crew discovered that the actual location of Bishop Rock was a mile from where it should be, a supposedly precise celestial waypoint James Alden had established shortly after MacRae's death. It was hypothesized that shockwaves from a powerful earthquake that had just struck Chile had somehow shifted the seafloor. It's more likely that Alden's reckoning was not quite dead.

A year later, *Los Angeles Times* writer George Wycherly Kirkman set out for the Bank aboard a fishing vessel. He was spellbound by the Bank's gin-clear waters and the possibilities of what lay below. Yet he said that fisherman reported the presence of more than "finny prey."

From unknown ports, they say, come silently sailing over these sunken isles weird phantom ships bound for harbors that are never reached by the ghostly crews that line the rails of their shadowy craft, dimly seen through the shifting many-colored mists of the sea—perhaps that once sailed from some distant forgotten port to which they never yet returned. Amid the half-lights 'twixt dark and dawn, weird voices are heard in quaint old sea chants in unremembered tongues, the clatter of ship gear, of slatting yards against creaking masts, and loudly shouted orders backed by blood-curdling deep-sea oaths once familiar to the bearded ghostly crews. At times, from the doomed caravels and galleons rise melancholy laments like the wild wailings heard aforetime near the Isle of the Lost Woman, storm-beaten San Nicolas Island to the leeward.

Into the mid-twentieth century, ghost ships and mortal fisherman were arriving in increasing numbers, often risking their lives to do so, and it's thus surprising that the Cortes Bank hasn't claimed more boats or souls than it has. There have, however, been at least a few sinkings and terrifyingly close calls. For instance, in November 1952, an eighty-foot purse seiner, the *El Capitan*, was cleaved in half by a sister fishing vessel while both worked a giant school of mackerel above the bank in the dead of night. Twelve terrified crewmen leapt into the ocean as the ship foundered, sinking in less than five minutes. All were rescued.

Fate was tempted again in 1957, when a flamboyant diver from Redondo Beach launched the only known major treasure hunting expedition to the Cortes Bank. His name was Mel Fisher.

To a great many, Mel Fisher needs no introduction. He was the son of a California chicken farmer who craved adventure and the spotlight, and early on decided that his future lay along the ocean floor. Newfangled aqualungs were making diving accessible to the masses, and in the mid-1950s Fisher and his wife, Dolores, opened a successful dive shop in Redondo Beach. Fisher became utterly addicted to the hunt, eventually moving to Florida to explore countless wrecks along the periphery of the Gulf Stream. His white whale was the *Nuestra Señora de Atocha*, a treasure ship that had foundered off Florida in 1622.

Some called Fisher a quixotic huckster and a shameless self-promoter with no appreciation for the cultural significance of the graves and live coral he scoured with a huge, seafloor vacuum. Yet Fisher paid a heavy price for his obsession with the *Atocha*, losing his son Dirk and his daughter-in-law in a 1975 dive off the Florida coast. Ten years to the day after Dirk's death, Fisher's team finally found their ship. Eventually, they unearthed more than $400 million in plundered Indian icons, jewelry, and bars of pure gold from the shattered hull of the *Atocha*—the richest undersea treasure ever discovered.

I once interviewed Fisher around 1989 and discussed the ongoing legal battles he faced in laying claim to the treasure. He was certain that courts would eventually find in his favor. When I asked how he was so sure, he spun his trademark phrase with a twinkle in his eye: "It's simple. Finders, keepers."

In 1957, Fisher sought to locate the wreck of the *Santa Rosa* atop Bishop Rock. He recruited a team that included a diving piano bar singer and a roller-skating instructor. He and his wife showed *Los Angeles Times* writer Lee Bastajian an array of equipment—an underwater bicycle, a high-tech sled that could detect metals, and an array of underwater cameras. Tantalizingly, Fisher also displayed bronze he claimed to have recovered from the Bank and a chart titled *Ye Olde Map of Reported Facts and Tales.* Just below a trident-clutching Neptune lay the *Santa Rosa.*

"Normally 50-foot waves break over the rock," Fisher told Bastajian. "Thus our departure will await a relative calm—20-foot waves."

When that "calm" window opened in January 1957, *Times* editors tapped a young reporter named George Beronius—a novice diver who had taken lessons at Fisher's school—and the treasure hunters headed out to sea.

I recently met George Beronius and his delightful wife, Eleanor, on a sunny October day in Santa Barbara Harbor. Now in his eighties, bright-eyed, and in possession of a wry sense of humor, Beronius was shocked when I showed him pictures of the giant waves the Bank is capable of delivering. "When we got out there, the water was calm and flat—just like a lake," he said. "To think what you're telling me with 40- to 70-foot waves. That would have scared the bejeezus out of me. If I'd known that, I'd never have gone."

Beronius joined twenty-three others at Newport Harbor aboard an aging seventy-five-foot charter fishing boat, *Via Jero*, on a drizzly morning. Beronius later wrote in the *Times*: "Everyone was in high spirits, brought on in part by the knowledge that each on board was to receive a share of the $700,000 gold treasure, just as soon as we could pick it up off the ocean bottom."

Fisher showed Beronius his map, claiming it came straight from an Acapulco museum. The young reporter was skeptical, but listened eagerly. Fisher then showed him the sled. Beronius told me: "It was really two-by-fours and some chicken wire with a glass meter on it that was supposed to detect metal. My impression was that his map—it wasn't some chewed up, weather-beaten piece of paper—was just an ordinary map with markings. I was actually under the impression that the whole thing was kind of a fraud or a lark. That was how the editors saw it, too. A bunch of kids going out with a crazy guy and a phony map. But everybody all kind of just went along with Mel."

Via Jero reached the Bank the next morning, anchoring alongside the clanging Bishop Rock buoy in a steady rain. The divers were soon leaping cheerfully into a placid ocean. Cliff Hanson, a retired speedboat racer, first tried out his own 150-pound motor-driven sled—which may have actually been a disarmed navy torpedo—that he called a "sub-glider." Unfortunately, the machine's nose shattered due to the depth pressure, rendering it inoperable. A few hours later, water pressure also shattered the glass of the viewfinder on Fisher's chicken-wire magnetometer.

The mission then degenerated into a wine-fueled lobster fest. "The water was so clear," said Beronius. "You could see the rocks and all the growth. There wasn't much agitation at all. We brought up the biggest damn lobsters you ever saw. That certainly did help mitigate the fact that we didn't find any treasure."

The divers hunted, drank, and feasted for two days, heading home at sunset on the second day. Beronius awoke at around 1 A.M., feeling a searing heat radiating from the bulkhead alongside his bunk. "I said, 'That ain't right,' and went to tell the captain."

When Captain Irving Chaffee opened the engine room hatch, a giant geyser of flame erupted. Beronius and the crew grabbed extinguishers, but the rubber hoses were so rotten, they disintegrated. Everyone formed a bucket brigade. "But we didn't even have buckets," he said. "Just dishpans from the galley."

As the fire neared a fifty-five-gallon fuel drum, the first mate suggested to the captain that it might be high time to call the Coast Guard. "The captain replied, 'Well, I suppose we could do that,'" said Beronius. "He called them. But when they asked, 'Do you need assistance?' He said, 'No, we're okay.' We were okay? We were on fire and we'd lost an engine. Turns out he didn't want them to know that he didn't have a charter license."

Moments later, the first mate volunteered to don a water-soaked wetsuit and a tooth-to-toe wrap of wet towels. He waded into the inferno and managed to heave water directly onto the flames. The fire was extinguished.

Via Jero continued on its remaining engine, but Beronius was spooked and unable to sleep. He marveled at schools of dolphins playing in the phosphorescent glow of the ship's bow wake and wondered what else might go wrong. At 4 A.M., the fire reignited. "We damn near burned down that time," he said. "If that boat had been gasoline- instead of diesel-powered, we would have blown sky high."

By this point, at least a few lights along Orange County's rural shoreline were visible. Beronius breathed a sigh of relief when the captain told him he was heading toward the green and red entrance lights to Newport Harbor directly ahead. But a quarter mile out, Beronius realized something looked funny—the lights were switching from red to green. He grabbed the binoculars. "It wasn't Newport Harbor," he said. "It was a stoplight."

Chaffee slammed his only functional engine into reverse and the boat slowly, painfully came to a halt within earshot of the crashing surf along the Pacific Coast Highway in downtown Laguna Beach—just shy of a nasty slab of barely submerged reef. Slowly and carefully, Chaffee turned north toward Newport Harbor.

"You know, I had never actually made the connection that this was *the* Mel Fisher until you brought it up to me," Beronius said. "Looking back and realizing that this guy found all this treasure in Florida—this was really the beginning of a big career for him. This is all really going to make a nice after-dinner story," he chuckled. "The whole thing—it was just a complete fiasco."

The mystery of whether treasure exists atop the Cortes Bank lingers today. Despite repeated inquiry, the determined silence of the Fisher family hasn't helped resolve the question. Maybe they found nothing, and would rather not discuss an embarrassing lark on the part of their old man. Or maybe, after a generation of legal wrangling over their past discoveries, they're simply sticking to the unofficial treasure hunter's law Mel shared with me in 1989: "Finders, keepers."

Given what Cortes Bank is most known for today—as one of the world's premier big wave surfing spots—one obvious question is: Who, exactly, was the first person to ride a wave above the Bishop Rock? Like most other "truths" about the Bank, this has proved hard to pin down.

The most widespread assumption was that Philip "Flippy" Hoffman must have surfed Cortes Bank while abalone diving in the 1950s, and if not Flippy, then perhaps his brother, Walter, or his hard-charging son, Marty. If not Marty, people said, maybe it was another surfing godfather, like Pat Curren or Jose Angel. Yet none appears to have surfed out there. I queried other editors and old-timers I knew, and chased frustrating dead-end leads, until finally, on the urging of *Surfer's Journal* publisher Steve Pezman, I tapped the encyclopedic mind of Mickey Muñoz. A diminutive man, Muñoz once charged Waimea Bay alongside Greg Noll and donned a woman's swimsuit to stunt double for Sandra Dee's *Gidget*. Today he's in his mid-seventies, but he's still a regular fixture in the lineups from Lower Trestles to San Onofre.

Mickey thought it possible that two hard-core California watermen might have surfed the Bank. The first was Pete Peterson. "He was an intrepid explorer," Muñoz said. "He was surfing out at San Nicolas Island before people would even go out there—before the navy got really entrenched. Another intrepid diver who had *cojones grandes* was Frank Donohue. He wrote for the *Santa Monica Outlook*, he was a movie stuntman, a commercial diver, a construction diver. That guy was so ballsy, he'd load his boats up with so much lobster and abalone that if he'd stopped, he would have sunk. He'd run across the border and unload by going by a pier and throwing off sacks till he was light enough that he could stop. I definitely would not have put it past him to have gone out to Cortes Bank to surf."

Unfortunately, both Petersen and Donohue died in recent years, and I've been unable to confirm any claim made on their behalf. Still, Muñoz's leads revealed a loose brotherhood of interwoven lives, through which the mystery has perhaps been solved. Like Flippy Hoffman, Muñoz told me I absolutely had to talk to a Hawaiian named Ilima Kalama. "Oh, and there's another guy you should speak with," he said. "You ever heard of Harrison Ealey?"

I met Harrison Ealey on a bright October day in Oceanside, California. Today, he's a fit senior with a tiny gray ponytail, blazing blue eyes, and a root canal with a lightning bolt. "I had one with a dolphin, too," he said with an impish laugh. "But I pulled it out flossing my teeth."

When Ealey was little, he spent his childhood in Dana Point, Laguna Beach, and an amazing little bungalow he still owns on a hillside several blocks from Oceanside's fishing pier. "It was a great place to grow up," he said. "My dad had a little sailboat, and I was an adventurous little kid. I'd sail down along the street and into the estuary when the winter rain opened it up. It was a great time to grow up, too. I mean, Laguna Beach was all seashells and dirt roads. We had a little skiff over there in Woods Cove—near Betty Davis's house. Her boyfriend and Flippy Hoffman taught me how to dive and take abalone. Dive down ten, fifteen feet with a speargun and pull out whatever you needed for groceries— abalone, white sea bass, and corvina."

Ealey showed me a quiver of battered surfboards, including a couple he surfed as a kid. He grew up surfing with Mickey Muñoz and Phil Edwards, one of the best surfer/shapers ever to emerge from California. Through the mid-1950s, they surfed the legendary point break known as Killer Dana (before the Army Corps of Engineers buried the wave). There were then so few surfers on the California coast that if you saw another car with boards on the roof, you pulled over and talked story. Often as not, you'd turn around and follow them to their destination.

In the late 1950s, Ealey became an avid sailor and began regularly crewing and later captaining on racing and pleasure vessels that plied the waters between California, Mexico, and Hawaii. In 1961, Phil Edwards joined Ealey in helping to ferry to Hawaii a big ninety-foot sailing yacht owned by the president of Matson shipping lines. Shortly after they disembarked on Oahu, filmmaker Bruce Brown (of *Endless Summer* fame) captured Edwards on the first documented ride through the cylindrical barrels of Oahu's infamous Banzai Pipeline—a wave up to then presumed unrideable. Pipeline would become the performance wave by which all others would be judged. A few miles away, Waimea Bay would simply become known as the biggest rideable wave of its day.

Ealey spent that winter surfing Waimea. He recalls dropping into mammoth waves alongside Butch Van Artsdalen and Greg Noll. As we talked, he pulled out a big photo of himself sketching into a Waimea bomb next to Buzzy Trent. "Waimea was just scary," he said. "You stay down, tumble, roll, break the surface, and there's so much foam just to get through. Then you have to go through another set or two. If you lose your board, you can't get in. That's how guys were killed. The current would take them down toward Haleiwa, and they'd find their bodies a few days later amid the coral."

When Ealey sailed to and from Hawaii or Mexico, he always made it a point to sail past Cortes Bank during the day. That way, he could avoid the Bank's shallowest reaches and see if big waves might be breaking above Bishop Rock.

"Sometimes we'd hit big patches of kelp and just go, 'Whoa,'" he said. "Sometimes you'd see this smoking thing out there. You knew it was big, and just to stay away. When the wind was blowing, you could see this fog cloud from the wave. With the bioluminescence it must have been a beautiful sight at night."

In the following summer of 1962, Ealey made a passage from Hawaii back to California with a Canadian couple. They reached the Bank in July. Ealey was surprised to find relatively smooth seas, a sizable long period southern hemisphere groundswell, and a beautiful breaking wave. The Canadians were game for a closer look.

"It was glassy," he said. "Otherwise we wouldn't have stopped. We anchored in a questionable spot, though, on the backside of the wave. There was a lot of kelp and rock. We didn't have fishfinders or anything that tells you the depth, and we didn't find sand. We rehooked and rehooked the anchor and finally caught on something down deep. Then we went snorkeling. The water was so clear. It was just beautiful down there."

Afterward, Ealey left the Canadians on the boat and decided to paddle his big board around to check out the chilly wave. It was inconsistent, and far from huge, at least by the standards to which he had become accustomed. "I'd been surfing Makaha and Waimea, so it didn't look scary. But then again, it wasn't fifty feet either. It looked like it was maybe eight feet. A good-size wave, well overhead. I used the boat to line up. I watched it and watched it before I tried to catch one."

With his big board and low-key confidence, Ealey stroked into his first wave. It was a long and remarkably fast righthand point break wave that deposited him safely in deep water before continuing on its way to California. He paddled back out, rode a few more, and when the tide began to stir the current, simply paddled back to the boat and pulled anchor.

"If I'd never surfed in Hawaii, I would have been scared to death out there," Ealey said. "It was the middle of nowhere. But it just wasn't *that big*. I mean, sure, I knew something could pop up. I knew it could get big. It was in the back of my head while I was surfing. That's why we didn't spend the night there. We went in, anchored, spent four or five hours, and split."

I asked Ealey why he never told anyone in the surf media about the episode— particularly after the wave burst into the consciousness of the surfing world three decades later. Ealey just shrugged. "I mean, I just wasn't all that associated with the surfing world. I kind of got away from the group when they became more financially interested in shirts, retail stores, clothing. They went one way and I went the other."

Ealey never stopped surfing or traveling. At one point while his boat was hauled out for repairs in Miami, he even befriended treasure hunter Mel Fisher.

"I never married either. No one could put up with my lifestyle. I just wasn't ready for the diaper service and the white picket fence. I mean sailing, surfing, it's exciting. It's an adventure. It's one of the few things left around where you're totally responsible for your own outcome—except for maybe going across the Mojave with a donkey and a canteen. Everything else is stoplights, or follow the yellow line, or do this but don't do that. Surfing's one of the few things left."

The circle of sailors, divers, and surfers was relatively small during Ealey's heyday from the 1950s into the 1970s. Smaller still was the troupe of hard-core watermen who included Cortes Bank in their perambulations, either to satisfy a jones for serious free diving or to provide a nice, if terrifying living spearfishing or harvesting abalone in the middle of the deep blue sea. One of these was an Aussie who once crewed with Harrison Ealey named Rex Bank. When Bank wasn't sailing, he would don a deep-sea drysuit and pry abalone off the seafloor off San Clemente Island, Catalina Island, and occasionally Cortes Bank. Six months of four-hour hunts for the big monovalves would leave him with enough money to sail or chase fresh powder in Colorado for the rest of the year.

One winter's day around 1967, Bank and a dive buddy named Kenny Cohen journeyed to Cortes aboard Cohen's twenty-eight-foot abalone boat. "I've been a surfer and a diver and consider myself a waterman," a ruddy and well-weathered Bank told me recently during an interview at his home in Long Beach. "I lived on the North Shore for a couple of years. That was terrifying. But to *surf* Cortes? I mean, those guys have got some balls. Diving out there, it was stupid. The place just always struck me with fear. There was a guy named Larry Doyle. He was a big guy, and he could really swim. One day he decided to go diving—to see what it was like under those huge waves. He just got beat to shit."

Bank recalled his own near-death experiences. "You go down these sixty-foot-deep canyons where the abalone live," he said. "One day, I was walking on the bottom and a big white shark appears. There I was blowing bubbles, and sitting down there in this bright green suit completely helpless. I was shit scared. When he turned, I climbed the hose and got the hell out of there. God, it was terrifying."

One night after a solid haul of sixty dozen reds and pinks, the men anchored above the Bishop Rock alongside another boat beneath a stellar full moon in sixty feet of water. The air was utterly still, the light off the ocean wondrous. Bank turned in, content in the knowledge that they'd have another banner take tomorrow and a fat paycheck. Some hours later, he awoke to feel the boat free-falling through the air. She slammed onto the ocean's surface, and a crash like artillery fire exploded from astern. Bank ran on deck to find a giant black monolith looming above her bow—a wave more than 40 feet high. It was the most terrifying thing he had ever seen in his years at sea. "It came out of absolute nowhere," he said.

The boat climbed the wave for an eternal few seconds, reaching near vertical, before yet another free fall. Cohen fired the engines while Bank yelled to the skipper of the other boat. "Guys, get up, get up! We've gotta get out of here."

Yet no one stirred as another wave, twice the length of Cohen's boat, carried them up its face. Bank hurled a ten-dollar abalone through the window of the neighboring boat, and her crew finally awoke, then screamed. Bank sliced the anchor line while Cohen turned the boat 180 degrees and ran like hell, the neighboring boat in their wake, a stampede at their heels. "The foam from the broken waves, it was above the top of the boat," Bank said. "We couldn't even see."

After they cleared the foam and collected their thoughts, Bank and Cohen realized the waves were only breaking atop the Bishop Rock. They wanted to see them closer. "The full moon," said Bank. "The calm air. The wave. It was just, just *beautiful.*"

Bank continued: "Kenny later shot himself to death in Sun Valley. The man was crazy. He had a death wish. I don't. I never went back to Cortes Bank."

Kenny Cohen took at least one more waterman out for his very last dive above the Cortes Bank—a compact, powerfully built Hawaiian with a perennial smile on his face named Ilima Kalama. Like Rex Bank's, his story is a lesson in how a big wave surf spot builds its reputation—by humbling the toughest characters in a small world built on pride and guts.

Kalama was born on Oahu and now lives on Maui. He comes from a long line of Hawaiian heavy watermen. His son Dave is a highly regarded surfer who for a decade and a half was the low-key towsurfing partner of big wave legend Laird Hamilton. Ilima's father, Noah, was also a renowned waterman who, in 1958, loaded up the family and moved to Newport Beach, California, bringing the sport of outrigger canoe racing to the mainland. Ilima rabidly surfed Orange County's chilly water through high school and won the West Coast Surfing Championships in 1962.

In 1963, while working as a lifeguard, Ilima was asked to pen the liner notes to a genre-defining album by the Ventures, a work of blazing guitars simply titled *Surfing.*

> Surfing is more than just a sport . . . it's a fever. Surfing has become a state of mind . . . a wild, uninhibited existence that revolves around the sun, the surf, and the sand. . . . Plummeting down a hill of moving green water and being able to move your board right or left—up or down—is a feeling

akin to flying, to skiing, and to sailing. The difference in surfing is that not only are you moving, but the force you've harnessed is also moving. . . .

The life of a surfer has a definite rhythm and beat to it . . . the beat of the surf and the beat of the wild, driving music he listens to. More than any other group, the Ventures have this sound . . . the beat of the surfer and the sound that he associates with the driving ride through the curl . . . a wipe-out . . . the life of the beach.

In 1966, Kalama joined a big, solid waterman named Larry Doyle on his first abalone dive out to the Cortes Bank. "I met some friends who were ab diving out of San Pedro," he said. "Hawaiian boys—former black coral divers. Oh, there were lots of pinks and reds at Cortes Bank. They were fetching about five dollars a pound, and there was lots of money to be made. The diving out there was amazing. I mostly remember how the water was just the most beautiful blue. Along the bottom it was rocks and sand and where we mostly dove, the surface was fairly flat and regular. There was no big tree kelp, more just small palm kelp. We went out probably three to four times on calm days and had good experiences."

On a placid morning in early 1971, Kalama climbed aboard the *Sea Way*, Larry Doyle's twenty-seven-foot cabin cruiser. When they reached Cortes Bank, they found four other abalone boats scattered around, and set anchor in thirty-five feet of water with the clanging Bishop Rock bell buoy in the near distance. It was still daylight, so they both decided to climb into their gear and dive. "We found a hot spot," Kalama said. "I brought up twenty-four or twenty-five dozen. Larry made a dive and came up with about the same."

Around midnight, Doyle awoke Kalama to tell him that the seas were getting rough. Other boats were leaving and Doyle thought it prudent to do the same. Kalama thought otherwise. He told Doyle that despite steep wind waves that were lapping over the boat's bow, they should just wait till the next morning so they could lever another motherlode off the rocks.

"Five hours later, he wakes me up again and says, 'Ilima, the boat's sinking.' The water's up to my knees. I told him to call the Coast Guard and to tell them that we're at the Bank, that we're sinking, and that we were going to try to swim to the bell buoy."

A few minutes later, the *Sea Way* disappeared beneath its fully clothed crew, leaving them to swim in breathtakingly cold water that Kalama reckons hovered in the high forties—a temperature capable of inducing hypothermia in a matter of minutes. Kalama was so disoriented that it took a while to realize that the reason he could barely swim was because he was still wearing his boots. He kicked them off. Then, after a few minutes in the water, as things seemed utterly bleak, came the epitome of a miracle. Both men's thick, hooded wetsuits unexpectedly corked to the surface like Queequeg's coffin in *Moby Dick*. They shed their clothes and climbed in.

The men next made a desperate swim for the blinking Bishop Rock buoy, but the black sea was a combat zone of steep, short-interval 12-foot swells, smothering whitecaps, and strong currents. They were rapidly swept out into the open ocean. "I've been out in the water in big waves," Kalama said. "Makaha and Sunset. I've been near death and near drowning, but none of that was even close to this. Larry was swimming right next to me. I told him to relax, and after a few minutes I started trying to crack jokes. We were drifting toward San Clemente Island."

Kalama didn't know it, but the situation was still beyond grim. In his near panic, Doyle had forgotten to give the boat's name or identifying number in his mayday broadcast. The Coast Guard ignored the call. The light of morning revealed a sight of terrifying desolation. The other boats were gone. The men were tiny, insignificant creatures, treading alone atop a vast ocean.

Kalama thought about his family—his wife, young son, and daughter—wondering if God would ever grant him the privilege of laying eyes on them again. He directed his prayers toward his mother, who had recently passed away. "My mom—I felt she was out there with me," he continued, his voice breaking. "That *mana*. That spirit. I felt so at ease. It was sort of, *what will be, will be.*"

They bobbed alone for perhaps a couple of hours. Then in the far distance, they saw a lone fishing boat. It appeared to be motoring slowly but generally in their direction. It disappeared in the troughs of the steep swells, but when boat and diver were simultaneously borne to a wave's crest, it would reappear, each time slightly closer. After an hour or so, she was perhaps a mile off. Then, unexpectedly, she veered away. "Our hearts just dropped," Kalama said.

Kalama and Doyle reached a decision. They would attempt the forty-mile swim to San Clemente Island. The effort itself would offer them some sort of hope—something to focus on. Yet deep down, Kalama knew this was essentially impossible. He had canoe-paddled the brutal twenty miles from the mainland to Catalina Island, and the relentless twenty-six miles of wind and current that separated the Hawaiian islands of Molokai and Oahu. What were the odds of *swimming* forty miles to San Clemente? But if they continued simply treading water, slowly bleeding heat into the Pacific, their chances were zero.

Then the fishing vessel reappeared to the south. Her captain had chanced to have his radio tuned to Doyle's channel during the mayday. He recognized the panic in the mariner's voice and proceeded toward Bishop Rock, where he eventually sighted a small but telling oil slick and, shortly thereafter, a floating hatch cover. He commenced a zigzag transect of the waters. A couple of hours later, he was shocked when he literally stumbled upon two men bobbing in mile-deep water. Their death march had ended just in time for lunch.

A month or so later, Kalama did something that I've found an astonishing hallmark of many who've nearly died at the Cortes Bank. He simply couldn't ignore the sirens—and went back.

"After we got saved and came back in, I was able to get on another boat," Kalama said. "I wanted to dive again at the same spot—because if I'm going to dive again, I wanted to do it where we sank. I went with another fella, Kenny Cohen. When we got out there, we saw that the waves were up. Cortes was breaking—the first time I'd seen it break. It was a good 25 feet, perfect rights. *Perfect*. I wish I had friends out there with me to go surf. It was just an amazing sight."

Because of the current and turbulence the swell was bringing in the shallower water, it was impossible to free dive at the usual depth of thirty to thirty-five feet. They dove without scuba tanks down to ninety or a hundred feet, where the currents were somewhat manageable. "All that time, when I came up, I could see the waves breaking just five hundred yards away," Kalama said. "When we were done that afternoon, we anchored, and wouldn't you know it, at one or two in the morning, the wind and ocean got rough. I pulled anchor and the ocean was almost washing over the back of the boat. It was the slowest we ever went to San Clemente Island. Kenny and me were just getting pounded left and right. You can't believe how many Lord's prayers I said on that trip. After that, I hung it up. I said to myself, *I think Akua wants me to stay on land*."

As we spoke, Kalama grew quiet for a moment. He had, it seemed, been granted freedom from the belly of the whale not once, but twice. "You know, I grew up respecting the ocean. I was taught by the very best watermen in the world. When my buddy woke me up that night at midnight, we should have left. It was greed that made me want to stay longer. I totally disrespected myself and disrespected the ocean, and I should have known better. I just thank God that in spite of that wrong choice that I'm here to talk about it. Thank God. *Mahalo Ke Akua.* To this day, no matter what I do, I cannot thank Him enough for the goodness and happiness I've received my whole life. I'm so thankful and so blessed. No matter what I try to do, I can never give back enough."

When Ilima Kalama speaks of greed and respect, and the fatal dangers that can follow an unwise decision, he captures an essential element of life atop the Cortes Bank. Bishop Rock's litter of wrecked boats, Larry Doyle's among them, are testament to this. Yet two far larger ships have also met the Rock. Both misadventures, one a near disaster and one an outright fiasco, would only add to the Bank's ominous legend.

Next to the USS *Constitution*, the ship that first discovered the Cortes Bank, the USS *Enterprise* is the second-oldest commissioned ship in the U.S. Navy. The aircraft carrier first sailed in 1961, and at 1,123 feet (about 300 feet shorter than the Empire State Building), she remains the longest U.S. Navy ship ever put to sea. She is also incredibly fast, her reactors still capable of pushing her along at forty miles per hour.

The *Enterprise* has, of course, seen duty all over the world. She blockaded ports during the Cuban Missile Crisis, served as a base for countless sorties over Vietnam, and in 2001, she was one of the first ships to launch airstrikes against Al Qaeda. Yet in 1985, she was very nearly undone by Bishop Rock.

At that time, her captain was Robert J. Leuschner Jr., a local boy who grew up along the paradisiacal shores of San Diego. He vividly recalls boyhood hunts for yellowtail skipjack around the offshore Coronado Islands. "It was always a race to see if we could get it in the boat before hungry hammerhead showed up," he told me. "The sportfishing fleet also ran occasional overnight trips to what they called 'sixty-mile banks.' I always wanted to go, but couldn't come up with the fare. Ironically, I got there thirty-three years later."

On November 2, 1985, a mere month before Larry "Flame" Moore first painted a bull's-eye on Bishop Rock, fifty-year-old Captain Leuschner stood on the bridge of the *Enterprise*, in charge of a five-thousand-man crew and a busy, swaying one-runway airport that resembled a floating city. The *Enterprise* was conducting an Operational Readiness Exercise (ORE) west of San Clemente Island that would bring her to a simulated "choke point"—an imagined tight passageway just beyond the Cortes Bank at 6 P.M. OREs are intense, fast-paced combat simulations, and Leuschner had expressed concern to his commanders that his flight crew was not yet ready for this "graduation day" level operation, but it was ordered anyway. Planes were to be launched in rapid succession while a myriad of other ship-wide drills were conducted—a man-overboard recovery being among the first. Drills and discussions pulled Leuschner off the bridge for extended periods while bad winds and course corrections eventually left the *Enterprise* two hours late for her choke point rendezvous. She accelerated to thirty knots and was bound for the Bank by 5 P.M.

At around 5:25 P.M., Leuschner returned to the bridge to note that an officer on duty had put the ship on a near 180-degree southerly course, heeding the navigator who intended to avoid the Bishop Rock. "That's dumb," Leuschner told the cowed officer, and he ordered a northerly turn to 322 degrees. Leuschner was annoyed. The southerly course would not have allowed for proper winds to recover three aircraft still in flight, and he was currently the only man on the bridge qualified to solve the wind, and swell-direction issues to reel his planes back in. Today he admits that he also became distracted by other tactical discussions, and thus missed a simple mention by the navigator that the *Enterprise* was nearing Bishop Rock. By the time Leuschner changed the ship's direction, *Enterprise* had already crossed a deeper, southerly reach of the Cortes Bank.

Naturally, today the *Enterprise* would have held GPS and/or sonar to warn her of shoal water. But in 1985 the GPS satellite array was not yet operational, and carriers were not typically equipped with sonar. Navigation was conduced with LORAN and charts. "I doubt if anyone ever thought of sonar as a grounding prevention system," Leuschner said. "That task was rightfully left to the COs [commanding officers like himself] and the navigators."

By 5:30 P.M., the *Enterprise* was ominously paralleling the outer edge of the Bank in the autumn dark. Leuschner noted white floodlights about four nautical miles off her bow. He correctly presumed it a gathering of fishing boats, but incorrectly reasoned that a dimmer, flashing red light was one boat's net buoy marker. Leuschner tried to hail the fishermen to inform them of his position, but received no response. A new navigator, meantime, had only recently begun his shift and began anxiously trying to single out the radar signature of the CB1 (Cortes Bank) buoy on Bishop Rock amid the cluster of boats. On deck, a plane attempting to land missed the deck arresting hook and was forced to make a nerve-wracking second pass.

Disaster might still have been avoided had an alarming call not come in to the bridge at 5:35 P.M. A senior officer belowdecks reported that a man was walking around with a 9mm Uzi submachine gun, and for a critical twelve minutes he kept the junior operator on duty (OOD) on the phone trying to sort out the security issue. "This might have been appropriate on a small ship where the OOD is the action officer on all abnormal events, but not a five-thousand-population carrier," said Leuschner.

Leuschner was consumed with airplanes, fishing boats, and a series of other decisions. A further underling, who knew the ship's location, might have warned the OOD to get the hell off the phone, but he was simply too scared to do so. Finally, Leuschner decided to look at the bottom contours that had apparently lured the small scrum of fishermen. When the navigator told Leuschner he would "like to give the Bishop Rock a wider berth than a thousand yards," the captain was stunned.

It was too late to turn the hulking machine hard left into deeper water, and Leuschner immediately ordered the *Enterprise* turned hard to the right, to a near due-northerly course, knowing she was likely already in water less than a hundred feet deep. He figured they were still a mile off the buoy, but when the ship reached a sharper angle to the Bishop Rock buoy light, he realized how close they really were. Moments later the immense aircraft carrier vibrated like a car passing over a series of washboard bumps.

The *Enterprise* struck the Bishop Rock a glancing bow, passing through a saddle-back formation and tearing a sixty-foot gash in her port hull. She began taking on water rapidly and was soon heeled over in an eleven-degree list. Leuschner ordered her counter-flooded to starboard to bring her off the list, while damage assessments were taken belowdecks. The next day, a contingent of marines leaned over the flight deck with machine guns, ready to blast any sharks that might threaten divers inspecting the hull. The divers found the gash, a ripped-off port keel, and severely deformed outboard port propeller blades. The tear would introduce jet fuel into the ship's drinking water supply, but the inner hull was miraculously unbreached. Despite $17 million in damages, the *Enterprise* was, remarkably, able to continue her exercise.

The Uzi would turn out to be a convincing fake, maybe even a water gun, and ultimately, it was perhaps the final of many contributing factors to what was an extremely unusual grounding of an aircraft carrier in any type of water, much less the open ocean. Indeed, the only other time the *Enterprise* ever touched bottom was a year earlier in San Francisco Harbor, when she scraped across a sandbar under the command of another captain (and oddly enough, carrying actor George Takei, crewman of TV's starship *Enterprise*, on board as a guest).

Admiral Leuschner called his own grounding a classic and highly dangerous "communications breakdown." Then he added, "But none of that diminishes the unambiguous responsibility of the CO [himself] for all outcomes. That's navy tradition, with which I wholeheartedly agree."

Leuschner would lose his command over the incident. He would go on to serve four years as a rear admiral, overseeing the development of computer-based combat systems before retiring. Yet Leuschner is under no misconception that things might have turned out far worse, for the ship and his crew, had the *Enterprise* struck Bishop Rock in even shallower water. Some might call it luck that it did not, but echoing Ilima Kalama, Leuschner said, "I prefer to think of it as God's will that no one was injured."

Robert Leuschner and Ilima Kalama represent two men humbled by the Cortes Bank who believe God spared them from worse fates. Yet the bones of one great ship do lie atop Bishop Rock—a ship, actually, larger than almost any that has ever foundered off the coast of California. How the SS *Jalisco* came to lie beneath three feet of water is another lesson of hubris and utter obsession brought on by the Bank and struck down by the hand of the almighty. It's also one of the most bizarre and harrowing stories of maritime survival I have ever encountered.

Chapter 4:

THE KINGS

OF

ABALONIA

It seldom occurs that new islands arise out of the sea.
But if it should happen that a new island arise, we state that
it must belong, as property, to whomever inhabit it first. But
he or they who colonize it owe obedience to the lord within
whose dominion the new island arose.

—Alfonse the Wise of Castille, 1265

On Halloween 1966, a story headlined "Pair Planning Island Nation off San Diego" appeared on page 11 of the *Long Beach Press-Telegram*. A small team of California entrepreneurs had bought a huge surplus navy freighter and planned to scuttle her atop the Cortes Bank in very shallow water. Their plan was to turn the Bank back into an island for the first time in thousands of years and then claim jurisdiction over its incredibly productive territorial waters. The men were, in short, nation builders.

What's amazing is not that these founding fathers failed, but how close they came to succeeding.

It wasn't until the second engine aboard *Sallytender* died that Joe Kirkwood began to truly fear for his life.

For the preceding hour, a twenty-six-foot fishing vessel and her ill-prepared two-man crew had been assaulted by chaotic 15-foot seas spawned by a relentless Santa Ana gale. Neither Kirkwood nor his buddy, Dick Hall, had any business venturing so far offshore. Death now seemed likely.

It was February 10, 1966, and the trip was meant to be a scouting mission to locate and simply get a feel for the waters around the submerged seamount Kirkwood planned to resurrect as Cortes Island. Not wholly unlike their ancient forebears the Kinkipar, Kirkwood and Hall had motored out past San Clemente Island late the night before. Yet, knowing nothing of celestial navigation or of following the invisible lines painted in the sky by seabirds heading toward a source of plentiful food, these pioneers had put their faith in technology, hoping to stumble upon the lighted buoy atop Cortes Bank and then spend the ensuing day leisurely snorkeling and exploring the waters around Archibald MacRae's Rock. Yet the night passed without any sign of the buoy, and with daybreak, its flashing light disappeared, making their quest akin to finding a needle in a haystack. They scoured the trackless sea from dawn to dusk, and then turned back toward Newport just as the wind began to howl. Seas went to hell in a heartbeat, and the *Sallytender* was battered by an ocean blitzed with steep, angry swells.

One of her engines died, followed minutes later by the other. She turned broadside and pendulumed nearly 180 degrees in the precipitous, short-interval waves. Kirkwood and Hall gripped their seats like bronco riders but they snapped off their mounts. Kirkwood radioed the Coast Guard, who explained it would be six hellish hours before the cutter *Point Divide* could reach them. The men were to check in every twenty minutes so a radio bead could be maintained on their position, a feat greatly complicated when *Sallytender* began taking on water. The Coast Guard dispatcher's voice rose a notch on this news. A helicopter would be sent.

The men bailed and reeled wildly across the cabin. At some point, Kirkwood, half-dead with exhaustion and seasickness, looked up to see Hall stuffing his face—a full stick of butter, a loaf of bread, anything in the galley. "How the hell can you eat at a time like this?" he asked. Hall replied through an overflowing mouth: "Whoever eats the most will last longest in the water."

"That was cool thinking," thought Kirkwood. "I couldn't argue with the wisdom of it."

Describing the moment later, Kirkwood wrote, "It almost came as a shock when I realized that the good physical condition I'd always prided myself in didn't mean very much out there. The sea was foreign to me and I found it to be a savage environment; completely different from anything I'd ever known, and one in which man is nothing. The slamming and the thudding and heaving were never ending. We were jarred to the teeth every few seconds. . . . We had quickly been exhausted, but the sea, far from being spent, increased in force almost as though to show us how puny, how insignificant we were.

"What the hell was I doing out here anyway? Only an idiot would attempt such a preposterous scheme as building a new country; and only a numbskull would be out here trying to do it. But even as the thought took form in my mind, while the numbing exhaustion crept through my mind and body, I knew it wasn't true. I didn't mean it at all. Building this country was exactly what I wanted to do—what I *had* to do. And my being there, being part of it from the inception, to watch it grow and take shape, from dream to reality, was just me. That was my way. Otherwise there was no joy in it."

After another endless hour, a Coast Guard chopper appeared on the horizon and Hall and Kirkwood sobbed in relief. They were safe, but the near miss would leave Kirkwood with a stark realization: It would take far more than a twenty-six-foot boat, a loaf of bread, and a stick of butter to conquer the Cortes Bank.

He knew exactly who to call.

For James Houtz, the mid-1960s were heady days. In early 1965, the former navy demolitions expert had set a world cave-diving depth record, descending to an astonishing 315 feet with a scuba tank filled not with an exotic mixture of nitrogen or helium but simple compressed air. The accomplishment was remarkable not only because it had bested a Jacques Cousteau record by a hundred feet but had been performed in the claustrophobic depths of Devil's Hole, an abyssal deepwater cave whose only access point is a tiny gash in the earth near Death Valley.

The expedition had earned Houtz considerable renown and a grant for further exploration. He now owned his own dive shop, was making a living doing what he loved, and was a happily married father to boot. Life was good.

Yet if Houtz was already well known at the start of 1965, a rescue attempt five months later very nearly turned him into a household name. Four young friends had descended into Devil's Hole. When two failed to surface, Houtz was called to lead the rescue.

Houtz and his team twisted and turned through labyrinthine limestone passageways hoping to find a terrified diver dog-paddling in one of the cave's primordial air pockets. Yet all he ever discovered was a scuba mask. "They were never found," he said. "The only thing down there now is bones."

All of which is to say that when, a month later, the phone rang in his dive shop, he understood the correlation between foolhardiness and death, particularly in the water.

"Hello, Mr. Houtz," said a robust voice on the other end of the line. "This is Joe Kirkwood."

The caller needed no introduction. In the previous decade, the tall, dark, and handsome Kirkwood had starred in a series of films and ABC television shows, playing the real-life rendition of a friendly and soft-spoken comic strip boxer named Joe Palooka. He was the son of Joe Kirkwood Sr., arguably the best trick-shot golfer who ever lived. In 1948, the men became the first father-son team to win a spot in the U.S. Open. A year later, Junior accepted the first of at least four invitations he would receive to play the Masters; he finished an impressive seventh.

By the time he phoned Houtz, Kirkwood was no longer acting or golfing professionally. He owned a bowling alley in Studio City. Shortly after the initial call, Houtz drove up to meet a man he found a charismatic wellspring of irrepressible energy. When the former celebrity boxer sat back in his chair and asked, "Have you ever heard of a spot called the Cortes Bank?" Houtz's brows arched. He had pulled thirty-five-pound lobsters from deep caves atop Bishop Rock, and he had fished among swarms of feeding albacore so frenzied that the water boiled for miles. An hour amid a school like that, and your forty-three-foot-long sportfisher was stuffed to the gunwales with high-grade sashimi. Yeah, he was familiar with the Bank.

Kirkwood announced his intention to refloat Cortes Island, and Houtz's jaw slackened. The effort was already well underway. Kirkwood had lined up solid financial backers and a partner with a huge rock quarry in Ensenada. He had signed a memorandum of agreement with the Los Angeles sanitation district to barge three thousand tons of landfill garbage to the Bank *a day*. There was money and legend to be made. What Kirkwood needed was a partner with Houtz's diverse maritime skill set to survey the Bank and help figure out how best to proceed. Houtz immediately thought Kirkwood "was nuts," but he was also clearly a big thinker and a risk taker. Houtz had spent his life among men of his ilk.

"The idea really got the wheels spinning," Houtz said. "It was like, 'Okay, I've got a project here. What will it take to do it?' My philosophy is, I don't accept the word impossible—I *can't*."

He would have to convince a highly skeptical wife, and take out a second mortgage on the family home, but Houtz was in.

Today, Jim Houtz is in his seventies. He's fit and compact with a pair of blazing blue eyes, a mischievous grin, and a few wicked scars on his arms and legs from motorcycle racing and other treasured memories of an ill-spent youth. He recounted his tale on a flawless fall day at the immaculate Laguna Beach hilltop ranch house he shares with his lovely wife, Joan. From his backyard, the Pacific is an endless cobalt expanse that stretches easily to San Clemente Island. With a telescope, it's easy to imagine you might see waves above Bishop Rock.

When I contacted him, Houtz was surprised that someone wanted to hear of his experience atop the Cortes Bank. He was even more surprised when I showed him a sixty-page manuscript that Joe Kirkwood had written about his role in the adventure. The manuscript and a small collection of photos had reached me anonymously and out of the blue. It seems Kirkwood penned the tale around 1967 and sent it to *Sports Illustrated*, whose editors surely knew of Kirkwood from his golf career. For some reason, it was never published.

"His story—it's in thirds," Houtz said, flipping through the pages. "One-third is fact. One third is a theatrical script for a movie. One third is fantasy—over and above all to cover his rear end. I will say, if this *had* been published anywhere, I'd have sued his ass."

Thus, what follows is a hybrid, a combined tale of sometimes competing perspectives. Only two men, Houtz and Kirkwood, really know what happened atop Cortes Bank, and Kirkwood, it seems, is no longer around to counter Houtz's differing recollections.

Shortly after his meeting with Kirkwood, Houtz learned of the other principals in the operation. Tony Aleman was the son of a former Mexican president. Robert Lynch was the president of a savings and loan. Bruce McMahan, the heir to a chain of furniture stores, owned a fleet of abalone boats and a rock quarry in Ensenada. When it came time for the men to name their new nation, Kirkwood liked the ring of "Lemuria," after a long-rumored lost continent of the Pacific said to have disappeared in a cataclysm thirteen thousand years ago. For some reason, though, the media latched onto the name "Abalonia," and this was the name that stuck.

Kirkwood's idea was simple. Dump enough of McMahan's boulders atop Bishop Rock and then sink a rock-filled barge atop that pile of rocks to return Cortes Bank to its previous life as an island. This would allow the Abalonians to plant a flag and establish a "monument," something akin to a mining claim, atop a shoal that, despite a U.S.-maintained warning buoy, clearly lay in international waters. The monument would quickly be surrounded with a growing donut of Mexican boulders and filled with a jelly center concocted of LA's landfill rubbish. Houtz met McMahan at a Tijuana watering hole to discuss how best to barge his rocks a hundred miles out to sea. "He was a little older than me—thirty-three, thirty-four. He seemed like kind of a spoiled kid. Nice enough, though. Where we met, they all knew him. All the chickees—everyone was around him."

Hours of brainstorming eventually led Houtz to a brilliant, seemingly simpler plan. Rather than scuttling a low-lying barge, it would be far easier to buy a big old ship, scuttle her on a level, shallow stretch of seafloor right around her actual waterline, and *immediately* surround her with rocks. Were the ship high enough out of the water to be inhabitable, her owners would create a more legally defensible "monument," along with a revenue-producing seafood factory from the get-go.

Eventually, with enough of McMahan's boulders and LA's trash, you could conceivably end up with a glittering seven-mile-long, three-mile-wide island resort atop the shoal's mesa-like reaches.

Kirkwood loved Houtz's idea. Heaven knows what the Kinkipar would have made of it.

Jim Houtz quickly found that the existing nautical charts created through the years seemed to record slightly different depths for Bishop Rock—perhaps lending some credence to Flippy Hoffman's rumor that it had once been dynamited. There were also simply too many gaps in the charts for a detailed picture of the bottom. Houtz would need to spend considerable time atop Bishop Rock—watching wind, waves, currents, and most of all, diving. By spring of 1966, he was making regular forays aboard his forty-three-foot, twin-engine sportfisher *Rainbow's End*.

During these trips, Houtz felt he witnessed nearly every sea the Cortes Bank was capable of dishing out—from swimming pool calm to blitzing winds and seas. He watched sizable waves crash above Bishop Rock, but nothing that might not be mitigated with a truly massive breakwater. He likened the approaching waves to a rolling barrel. The water actually only moves up and down while the barrel rolls. The key, then, was to shatter that barrel with McMahan's boulders.

The best spot to lay a ship seemed to be off the southwestern edge of Bishop Rock's two-and-a-half-fathom peak, about a third of a mile northwest of the buoy. There, the shallow ridgeline would have already dissipated considerable swell energy. The seafloor dropped precipitously off an adjacent ledge where fishing boats might make a safe approach to the ship's stern to offload fish, lobster, and abalone.

"We had a flat bottom—a beautiful bottom," said Houtz. "And it was sandy. We'd dig up these cockles—clams—eight, nine inches. Huge, absolutely enormous. Good God, did they make good chowder."

They found their ship—the one they would scuttle—in San Francisco. SS *Jalisco* was a mighty unusual vessel—the product of two facts of life during the years surrounding the great wars. First, the United States had an inexhaustible need for freighters. Second, steel was scarce, while concrete was relatively plentiful. The concept, then, was to create a ship-shaped lattice of steel rebar and surround it with a tough concrete called ferro-cement. The first of these concrete "Liberty" ships were rustproof, but slow, heavy, and fragile. In 1920, the concrete *Cape Fear* collided with the *City of Atlanta* at Narragansett Bay. She "shattered as if a teacup was hit," according to Rob Bender of the Web site Concreteships.org, sinking in three minutes and taking nineteen crewmen.

The SS *Richard Lewis Humphrey* was one of twenty-four identical concrete ships rushed into service during World War II. Each of these so-called McCloskey ships was 334 feet long and weighed five thousand tons. *Humphrey* reportedly carried a load of coffee to the Pacific Coast before being damaged in a storm and sold to Mexico for scrap. It seems evident that someone in Mexico instead rechristened her *Jalisco*, and she continued to ply the Pacific until sometime in the late 1950s or early 1960s. Houtz found her collecting rust and dust among the mothball fleet in Oakland's naval shipyard.

For eighty thousand dollars, the Abalonians purchased this serviceable vessel with a flat bottom and twenty-seven feet of elevation between her waterline and main deck. Her price included a workroom filled with industrial-grade metal and woodworking tools and a discount for salvage removal of her main engine, turbines, and other sundry parts. She would draw far less than her normal twenty-six feet with her weighty running gear removed, and would have to be towed down from San Francisco behind a tug. Houtz calculated that after scuttling, her lowest stretch of hull would be better than twenty feet above the high-tide waterline at Cortes Bank. This was good, for it meant she'd make for an immediately dry and habitable Abalonia.

As Kirkwood was handling most of the sale itself, Houtz gave him his list of requirements: The two forward holds must be insulated for seafood freezing and refrigeration. Two auxiliary turbine generators, 50,000 and 250,000 watts, and two boilers were to be fueled and online. And both massive anchors, their 750 feet of chain, and the diesel engine that drove the air compressor for their winches must be operational.

The sole issue that caused Houtz pause was *Jalisco*'s meager array of four ballast pipes—each only four inches in diameter. The ship needed to fill with water and settle to the bottom quickly, and these would not do the job. Houtz said, "Kirkwood told me, 'I'll add more valves.' I said, 'Fine.' I left all that stuff up to him."

Houtz knew leaving such important nautical requirements to an admitted landlubber might be a mistake, but he was too damned busy running his dive shop and worrying about logistics. He decided that the best chance for success would come in having *Rainbow's End* lay out a "runway" of buoys to direct *Jalisco* along and just to the north of Bishop Rock's shallow ridgeline. Once in position, Houtz would lock her into a precise position along his runway by letting her drift backward in the southerly current and playing out her anchor chains. Ballast valves would open and *Jalisco* would settle forever in thirty-two feet of water. A veritable conveyor belt of McMahan's boulders would quickly surround her.

Kirkwood tried repeatedly to take official channels to obtain some sort of blessing or approval from the United States for his island, but he was met by head scratching at the various agencies he queried. No one returned his calls. "Some people began to kid me, calling me 'King Joseph' or 'King Kirkwood,'" he wrote. "As a gag a friend handed me a fistful of rocks to 'help you build your kingdom.'"

As November arrived, Kirkwood's plans had become newspaper fodder, and they drew the attention of the city of San Diego, the U.S. Army Corps of Engineers, and U.S. Attorney Edwin Miller. There were valid concerns. What if Kirkwood was a communist sympathizer? What if the Abalonians decided to restrict fishermen in their newly claimed territorial waters? What if LA's garbage started washing ashore? What if the mafia wanted the island for a casino? Kirkwood claimed to have already refused such an offer.

Kirkwood became impatient, and then frantic, waiting for answers to questions that never came and fearful that the U.S. attorney or someone else in the government would order him to cease and desist. He phoned Houtz at the start of the second week of November. How was *Rainbow's End*? How was the weather? McMahan's rocks were ready. *Jalisco* was ready. It was time for action.

Houtz had just dropped a brand-new pair of Chrysler Hemi V8 engines into *Rainbow's End*. They needed at least fifty hours of break-in time before a long trip out to the Bank. But Kirkwood wanted to get this going *now*. Houtz studied the weather. A gale had wound up and pulled out to sea off Japan, but its wind and rain were many days out. His best guess was that the storm would track to the north, leaving a strong dome of calm high pressure anchored over Southern California. He reluctantly granted Kirkwood's request.

The team assembled at the Balboa Bay Club on the afternoon of Sunday, November 13, 1966. Kirkwood arrived regally clad: pleated khaki trousers, a nice sweater Houtz suspected was cashmere, and a pair of fur après-ski boots, which Kirkwood thought might keep his feet warm. Houtz clucks at the memory of the boots, their image permanently seared into his brain. "Everybody at the club had been looking at the boots and looking at me, and asking, who *is* this guy?"

The royal entourage included Houtz's navigator, a man whose name he has forgotten, and a fellow diver and employee he today only remembers as "Dan." Kirkwood brought along a pair of young men Houtz had never met—William "Many Horses" Lesslie, a short, muscular man of Native American descent, and a young assistant named John O'Malley.

With the King of Abalonia safely aboard, Houtz slid the throttles forward on the *Rainbow's End* and set a course for Cortes Bank. Somewhere off Catalina Island, an agitated Kirkwood asked Houtz to speed up. Houtz refused, saying, "Joe, look, we set up a plan. . . . You know the exact RPMs of the engine and the speed of the boat. That's how you know where you're going." As Houtz explained later, "It had been made clear on an earlier trip—this is my boat. I'm the captain. Stay out of the way."

An hour or so later, one engine emitted an earsplitting clatter. A flabbergasted Houtz ordered it shut down. He wanted to yell *I told you so* to Kirkwood, but bit his lip. The decision to make the run had, after all, been his. They would continue on minus an engine.

Houtz wondered why Kirkwood was in such a hurry, not realizing that a race between Kirkwood and the U.S. attorney had already commenced.

Rainbow's End reached Bishop Rock just after dawn on Monday morning. The weather was a California dream, the water as calm as a pond. Houtz took a bearing on the Bishop Rock buoy and located the spot for the first of his orange markers. "Everything was already charted," he said to me. "We had the lengths of line attached to the buoys with all the weights. All we had to do was drop one, stay on course, drop two, stay on course, drop three, stay on course."

A call came from the tug *E. Whitney Olsen*, which was hauling the *Jalisco* down from San Francisco. They would be visible on the horizon soon. Shortly thereafter, Bruce McMahan would reach Bishop Rock from San Diego aboard a sportfisher, the *Polaris II*. His boulder armada should arrive from Ensenada on Tuesday morning.

As Houtz laid the markers, the *Whitney Olsen* loomed off to the north, dwarfed by the silhouette of the *Jalisco*. She was a grand and eerie sight. Her hull was slate gray, but long rusty streaks ran down beneath patches where her rebar bones had been exposed—looking for all the world like dried blood. Toward sunset, Kirkwood joined McMahan aboard *Polaris* while Houtz remained aboard *Rainbow's End*. By nightfall, the *Whitney Olsen* and *Jalisco* would begin making the first of many broad, idle-speed circles well to the west of the Bishop Rock buoy.

Back on the mainland, the proceedings were being eyed closely by *San Diego Union Tribune* reporters, who possessed a radio telephone and a private channel to Bruce McMahan aboard *Polaris*. "People thought we were just kidding," McMahan had told them. "If all goes well, we should be starting operations out there in a couple of weeks."

Jim Houtz claims he doesn't know when the call came in, but at some point around dinnertime, a radio telephone conversation took place between U.S. Attorney Edwin Miller, *Whitney Olsen* Captain Cliff Miller, Kirkwood, and McMahan. The message was blunt. The Abalonians were on the U.S. continental shelf, they were in violation of U.S. federal laws, and they were to cease all operations.

Captain Miller was to await further instruction. Kirkwood wrote, "I gave Captain Miller the information to relay to the U.S. attorney regarding the people we were in contact with in Washington, and sat back to wait, chewing on my lip."

In about an hour, the U.S. attorney called back demanding the ship be towed to San Diego. "He then said we were in violation of U.S. laws because we were a hazard to shipping," wrote Kirkwood. "He commenced reading the law to us, but his voice seemed to falter as he said the words, '*misdemeanor, punishable by $50.00 fine.*'"

As Kirkwood recorded, "The captain, incredulous at the thought that we were a hazard to shipping, asked, 'On Bishop Rock?'"

Attorney Miller shouted back angrily, "Do you understand me, captain! Tow that boat back or I'll have your license!"

"I thought I had anticipated every angle," Kirkwood wrote. "But this was so illogical, I had considered it briefly and dismissed it from my mind months ago. We were on Bishop Rock, itself a great hazard to shipping, to which all the boats that had gone down there were testimony, but it was well protected by buoys put there by the United States and plainly marked on every map in existence. For that matter, I just couldn't see how the U.S. could have jurisdiction. But I had gambled and been fully aware of the risks. I tried to stifle the resentment I felt for the U.S. attorney and told myself he was just doing his job. He had only cited us with a misdemeanor. I could just ignore it, go ahead with the project, and pay my fines, but that would still be against the law, and that wasn't the way I wanted to do this."

But Houtz said that's not how it went down.

Houtz thinks a conversation then took place between Kirkwood, McMahan, and Captain Miller. U.S. attorney be damned, they were *going* to scuttle the *Jalisco*. But the right story had to be told. What if they claimed that *Whitney Olsen accidentally* scraped *Jalisco* across an uncharted portion of Bishop Rock *after* Attorney Miller's call, and she started taking on water? It would be impossible to tow a sinking ship to San Diego, so to avoid creating a *new* shipping hazard, they could claim they decided to sink it atop an existing shipping hazard, which conveniently lay right along Jim Houtz's runway. Captain Miller could keep his license, and the founding fathers could cry forgiveness, not ask permission, collect on a $45,000 insurance policy (or at least have the ship above the water so she could be salvaged), avoid criminal prosecution and a fifty-dollar fine, and *still* potentially witness the birth of their nation.

It was a brilliant plan. One that, Jim Houtz said, Kirkwood had clearly decided he wasn't going to share with Houtz. Because if Houtz balked and went along with the U.S. attorney—which he would have—Abalonia would never rise from the deep.

Joe Kirkwood and the captain of the *Whitney Olsen* would indeed claim that *Jalisco* struck the Bishop Rock sometime Monday night and began taking on water. Yet both Jim Houtz and *Whitney Olsen* crewman Louis Ribeiro, whom I interviewed recently in San Diego, insist that as dawn broke on Tuesday morning, a perfectly sound old ship stood ready to become an island.

From *Rainbow's End*, Houtz asked Clifford Miller to tow the *Jalisco* into position. She steamed in from some distance off Bishop Rock and lined up with the buoy runway. By 9:15 A.M., all was well.

The men marveled as the spooky old ship slid silently past *Rainbow's End*. Houtz pondered the origin of a round hole that had been punched in her forward gunwale—a nasty, toothy little wound about two feet around and lined with crumbling cement and twisted, rusty rebar. A lengthier section along her forward starboard bow had also shattered. She really was rotting to her bones.

Houtz planned to live aboard the ghost ship for the next few weeks. He grabbed a bag of supplies, a .270-caliber bolt action rifle (should he need to defend their island), and a life jacket, and he had his mate bring the *Rainbow's End* alongside the *Jalisco*. As best as Houtz can remember, Dan, Kirkwood, William Lesslie, and John O'Malley joined him in clambering aboard, ferrying life jackets and other supplies. Spirits were high.

It was at this point that Houtz first noted an odd sensation. The horizon was making a ponderously slow seesaw. *Jalisco* was undulating atop a very long, low, and pillow-soft swell that was approaching from her bow. I asked Houtz exactly how long, and he gave a whistle. "Long," he said. "The distance from trough to crest was tremendous."

Houtz had not seen swells like this out here, but he had also never stood right above Bishop Rock on a big ship. The swells weren't very big—two, maybe four feet. He and Dan discussed them briefly but shrugged and attributed the anomaly to conditions right atop the rock. "It's been dead calm," he told me. "So what if we get a little wind and the normal seas. I mean, this is a *freighter*. Big deal."

Houtz next made a basic walk around the ship to ensure that Kirkwood had *Jalisco* outfitted properly, per instructions. The first discrepancy left him thunderstruck. *Jalisco* held only one anchor, on her starboard side, and it was far smaller than those he had seen in Oakland. He hadn't noticed this from *Rainbow's End*. Where was the port anchor? Why was this one so small? Kirkwood umm'd and ahhh'd before admitting that he'd sold them for salvage.

The blood rose in Houtz's cheeks. Playing out the anchors precisely to maneuver *Jalisco* was now out of the question. Clifford Miller still controlled the ship from the deck of the *Whitney Olsen*, but Miller wanted *Jalisco* cast off as soon as possible. He had brought the ship this far and his work was done. Houtz thought a serviceable position might yet be obtained if he carefully positioned and played out the single anchor in the current. He keyed the starter of the eight-ton diesel that powered the chain spool compressor, but it wouldn't even turn over. Hadn't Kirkwood had it tested? "Nobody ever showed me that it worked," Houtz said Kirkwood replied.

There was now only one way to lower the anchor—a massive, manually operated bow winch. But, once that was lowered, there was no way in hell to raise it. Houtz ordered Kirkwood and his crew to slowly release the brake, but the chain raised such a horrendous clatter that the startled men leapt back and let go. It zippered out at lightning speed before coming to an abrupt, thumping stop.

A moment of stunned silence ensued. Houtz first thought someone must have managed to activate the brake because nowhere near 750 feet of chain had played out. Yet the spool had emptied. "Where's the rest of the chain?" Houtz asked Kirkwood.

Kirkwood shrugged. "He told the people at the shipyard, 'We're going to be in less than fifty feet of water,'" said Houtz. "He had the chain cut off and sold it for salvage to get more money—and he never said anything about it."

Houtz was furious at Kirkwood and himself, but argument or debate was pointless. They were in about 50 feet of water with 100, maybe 150 feet of anchor chain. Houtz assessed the situation, mentally calculating different decision trees. None bore much fruit. The only way the mission might be salvaged was if the *Jalisco* somehow ended up in a functional spot on the reef as the current pushed her backward into shallower water. Houtz ordered Kirkwood to open the ballast valves below deck so she would begin to sink.

Jalisco was now attached to an anchor that, while not yet latched into a hard, fast position, nonetheless lay on the seafloor, and this would make it impossible for the *Whitney Olsen* to drag her out to deeper water. The chain was so short that Houtz reckoned the anchor would probably grab bottom just as *Jalisco* drifted into a position astride the precarious ridgeline where he had seen waves break during the previous months. To make things even worse, *Jalisco* was still bound to the tugboat by the tug's heavy steel towing cable. Precious minutes passed as the men labored unsuccessfully to lift it off *Jalisco*'s bow stanchion. With the swells continuing to build, a fearful Captain Miller could bear no more.

Kirkwood hollered to Captain Miller to give the cable more slack. He was instead puzzled and alarmed when a man stepped out of *Whitney Olsen*'s cabin and fired a blowtorch. Kirkwood wrote, "The captain must be really worried by the increasing swells to cut away a cable worth several thousand dollars and let it sink in the ocean."

Whitney Olsen crewman Louis Ribeiro held onto fellow crewman Ray Turnbull while Turnbull torched the cable. Conditions were going to shit. "We were going underwater while he was cutting the thing," Ribeiro said.

From *Whitney Olsen* came a sharp crack and the Abalonians ducked. The melting cable had separated, whistling through the air with the speed of a striking cobra and ricocheting off *Jalisco*'s bow. She still floated freely in the shallow water, but not for long.

A building marine layer cast a funereal pall on the proceedings. With a sense of dread, Jim Houtz suddenly realized what was happening. The *Jalisco* was being enveloped by long, low-frequency forerunners that formed the leading edge of a big North Pacific swell. The waves had radiated out from the same low he had noticed days ago on the map, and they were now barreling down the California coast. The great old ship would soon be battling for her life at the mercy of the waves.

As the outer edge of the swell swept past *Jalisco*, wavelengths shortened into the twenty-second range, and the swells rapidly grew in surface height. Less than a mile off *Jalisco*'s bow, the swells encountered something they hadn't felt since Hawaii—an immovable obstacle. Their energy focused and compressed, but Bishop Rock wouldn't budge. The swells could go nowhere but up.

Jalisco climbed and then dropped sharply down the backsides of the waves, her hull ringing like a struck gong. Free of the tug, she was nudged backward. Her anchor scraped and bounced for a couple of hundred yards before finally grabbing hard in twenty, maybe thirty feet of water. The chain drew taut as a banjo string, and *Jalisco* shuddered violently, throwing the men off their feet.

Moments later, a 20-foot swell lifted *Jalisco*'s bow and a deep, rattling groan bellowed from astern. She had kissed Bishop Rock for the first time. Firmly tethered to the anchor, she was soon grinding and lurching against the ancient mountaintop—a stomach-clenching series of thunderclaps rolling through her hull. Houtz has still never experienced anything like it. "It was just, just the most god-awful thing," he said.

Off the bow, a new line of swells. The ship slammed the rock in the troughs and was suddenly shaken by a terrific concussion. The floor dropped from beneath everyone. "Whoosh," Houtz said. "It was like an elevator falling."

Jalisco had been fatally impaled, and the punctured portion of her stern instantly fell onto a pinnacle of Bishop Rock like a trailer on a ball hitch. The waves forced a tortured, deafening turn to port. With every inch, the pressure on the anchor chain grew.

"It's kind of like a movie camera that's gone into slow motion and now into still frame," said Houtz. "I told everybody to get away from the anchor chain because, when it goes, it's going to be something you do not want to be around. I came over to the port side and stayed there because that was the lee of the anchor."

Houtz was wearing a life jacket. At this point, he might have simply jumped overboard to save himself. But he was, officially, the captain of this sinking ship. It was his duty to help everyone get out alive.

When *Jalisco* reached a forty-five degree angle to the waves, the chain's weakest link split near the hull with another massive crack. It bullwhipped out across the water with enough force to cleave a boat in half. *Jalisco* gave a massive jolt and swung around with dizzying speed until she was stern first into the waves, roaring and gnashing against the stake that had been driven through her heart. As she sunk lower, the waves grew higher, not yet breaking but just beginning to wash over her backside.

The men stood just forward of amidship in a daze. Things had gone wrong so fast. Houtz said to everyone, "This thing is done, guys. We have to get the hell off this boat."

As the swells swept farther and farther over *Jalisco*, it became clear that the men were soon going to be cast involuntarily into a riot of white water if they didn't leap overboard first during a lull. Houtz ordered everyone into life jackets, but Kirkwood, Lesslie, and O'Malley refused. Houtz was flummoxed.

"I'm somebody who can swim a hundred miles, and I put on *my* life jacket," he said, shaking his head. "And they wouldn't do it. They wouldn't do *anything*."

I asked Houtz if this was an example of the "incredulity response." In the 2008 book *The Survivor's Club*—an examination of the traits found in those who survive utterly harrowing experiences—author Ben Sherwood writes that those who die in critical situations often don't believe what they're seeing and freeze like marble statues. Sherwood also describes a condition he calls "brainlock," when unhinged panic inhibits the ability to think your way out of a situation. A person might even do something completely irrational—walk in the direction of a fire or not put on a life jacket in the face of gigantic waves.

Houtz lit up. "That's *exactly* what was happening," he said. "I've had my life on the line quite a number of times. In each of those instances, it's not a panic. It's mostly a mode of 'Okay, what do I do now to get out of this?' But some people just don't listen. There's nothing you can do."

The first breaking wave roused the men from their deer-in-the-headlights stupor. When Houtz yelled "Run!" there was no argument.

The wave stood perhaps 35 feet high and was a hypnotic sapphire blue. It gathered up concrete, cast-iron hatch covers, wood, rope, steel cable, and fifty-five-gallon drums of diesel. The debris-field overtook everyone at the bow except for the speedy Kirkwood, who leapt a few feet up onto the base of a small mast planted on the very nose of the ship. The others were bashed against the bow gunwale and hammered by the debris. Houtz felt a stabbing, crunchy pain in his side as the breath was squeezed from his lungs. He had broken at least one rib.

A terrified John O'Malley had been battered, too, but "Many Horses" Lesslie was in real trouble. He had somehow been stuffed ass-first into the jagged hole in the bow and was bent over double, his body blocking the rushing water like a cork, completely unable to free himself. Another giant frigid wave swept the bow, and another. Lesslie was going to drown.

Then the set passed and there was a lull. Houtz and Dan staggered over and yanked out Lesslie, who was spitting out oily water and moaning in pain. Lesslie was indeed hurt, and O'Malley had probably suffered internal injuries. The pair then helped O'Malley into a life jacket and walked him to a spot just off the port bow. In a brave and foolish act, Captain Miller then rammed the *Whitney Olsen* into the *Jalisco*, revving the tug's engines hard to maintain contact with the hull. Houtz and Dan tried to hand O'Malley down to Louis Ribeiro and another crewman. The pair had him for a second, but O'Malley was covered with a slippery sheen of diesel. A wave pulled the tug down, and the young man "stepped out into the air," Ribeiro said. Miller withdrew his boat to fish him out.

"Get off the boat, Dan," Houtz said. "There's no reason for you to stay here." Dan leapt off the side and struggled over to the *Rainbow's End*.

Houtz moved to shelter himself behind the ship's three-story-tall superstructure, and from there he tried to talk Kirkwood and Lesslie down off the bow and convince them to jump into the water. Kirkwood shivered, clutching the mast in a death grip, while a weary, damaged Will Lesslie held on just below him in the sheltered lee of the anchor winch. The ocean had been ominously calm for a few minutes.

"I was yelling 'Joe, get back here!'" said Houtz. "He said, 'No. This thing's not going anywhere! I'll hold on and the water will just rush by me. It's gonna go by me.'"

The compressor had been lashed down, but one end of it had been broken loose by a wave, and it was just swinging around. If it had been broken free by a wave, it would slam Kirkwood and Lesslie like a runaway bulldozer. I said, 'Guys, come on. The compressor's coming loose. That thing weighs eight tons.'"

"I gave myself up to hanging onto that mast for all I was worth," wrote Kirkwood. "Absurdly, I was determined that no wave would wash me over, if for no other reason than that people are always being washed overboard in movies."

Houtz watched the water below the bow draw down. It was being gathered up by a wave. He peered around the corner at a thing of beautiful horror. The wave, the most massive he had ever seen from land or sea, stood high above the superstructure. Kirkwood gaped in wonder like Jonah before the whale. Looming above him, he wrote, was "an enormous wall of bluegreen water rising 45 feet or more, the fish in it plainly visible."

The wave roared down the deck. Green water exploded around Houtz as he stepped into the protected lee of the superstructure. Houtz managed one last, long look at a wide-eyed Kirkwood before the King of Abalonia was blown off the deck of his castle by the titanic fist of water.

"I remember seeing him just flying through the air," added Ribeiro.

Houtz had between fifteen and twenty seconds before the next wave. He peeked around the corner and gazed up at a sight he had previously thought impossible. This wave was 50 feet high—easy. When it slammed the superstructure, a dark ceiling erased the sky, a condition mariners call "green water." "You're looking way up and all you see is green water coming up and over that bridge," Houtz said. "It was solid water. Not spray, not a little bit of curl, just green."

A cubic yard of water weighs around 1,700 pounds—almost as much as a 1966 VW Beetle [at 1,672 pounds]. Thus, a mere fifty-cubic-foot segment of this wave weighed 7.75 million pounds, and smashed the superstructure at between 35 and 45 miles per hour. The roar was deafening, but the superstructure miraculously held. Ten seconds later, the maelstrom abated. Houtz ran to the railing to see a vast cauldron of seething water. Bucking and churning in the middle of it all was the *Whitney Olsen*. There was no sign of Kirkwood or Lesslie.

Of his experience, Kirkwood wrote: "Suddenly I was flying through the air with the mast still locked in my arms." Aboard the *Charger* (a boat that had arrived carrying several reporters), two Catholic newsmen went down on their knees and started praying. Something struck my head, a shattering blow, and I went down, sinking and drifting into blackness.

"I regained consciousness in the murkiness of deep water, already instinctively swimming and fighting my way up through the darkness, until at last I could see daylight above, and struggled to break the surface. I gasped and coughed up water, my voice rasping in my throat as I sucked in air, when I was suddenly hurled down through the water with the force of a building falling on me, deeper and deeper into the dark depths, the pressure on my ears almost unbearable. My lungs bursting for air, I again fought upwards until I reached the foam and emerged, treading water, fighting for a few seconds of air.

"There was a film of oil on the water and debris everywhere in sight. If I could only find something to hang onto! Something knocked against my arm, and I grabbed blindly for it, grasping onto a six-inch-long piece of two by four lumber. For a fraction of a second, I appreciated the irony. With half a ship floating on the ocean, I find a matchstick. . . . I tried to get my after-ski boots off. I had worn them for warmth, but now, waterlogged, they felt like lead on my feet. I tried to work the zippers down the front of them, but couldn't get at them without putting my head underwater. It flashed through my mind that I needed to lose a few pounds. I had just given it up when another wall of water fell on me, and I was hurled headlong down into deep water."

Anticipating a rescue, Cliff Miller had already staged the *Whitney Olsen* just off *Jalisco*'s bow. As near as Houtz can reckon, Kirkwood disappeared into the raging foam and was carried beneath the tug, the entire 120 feet from bow to stern. Mere feet separated the bottom of the tug from the top of Bishop Rock, yet Kirkwood passed safely not only between hull and reef but around two propellers. He boiled to the surface off the tug's stern and miraculously managed to weakly lift his head. Ribeiro stripped off his shirt and prepared to leap in for a rescue. "Are you crazy?" a crewman asked.

Ribeiro had spent his entire fisherman's life throwing lines. He instead heaved one out to Kirkwood with a skill Captain Miller compared to a big-league pitcher.

"No matter what you do, don't let go," Ribeiro yelled to Kirkwood, just before he was buried by another nightmarish wave. *That guy is one tough son of a bitch,* Ribeiro remembered thinking.

"I was flagging and knew I couldn't hang on any longer," Kirkwood wrote. "As one of the crewmen grabbed my hair and another my armpit, I blurted out, 'Please help me fellows. I can't help you.' And the rope started slipping from my grip. In seconds more they poured me like a sack of potatoes onto the deck, where I lay in a soaken, oily heap, too spent to move and beyond caring about anything."

Houtz looked around. Lesslie was gone, and he was all alone aboard the dying ship. He made sure his life jacket was secured tightly, clutched his gun, and stepped off the port rail before the next set of waves bore down on the ship. The current carried him toward the *Whitney Olsen*. He was plucked out of the water. Miraculously, so was Lesslie.

After several more sets of waves, the *Jalisco*'s entire superstructure tore completely free of the deck in a colossal mingling of water and steel. Anyone forward of it would have been crushed to powder.

In the backyard of his Laguna Beach home, Houtz flashed back to Kirkwood's last moment aboard *Jalisco* and shook his head. The pictures taken that day were shot by a former U.S. Marine combat photographer named Daniel Bresler, who worked in a photography studio across from Kirkwood's bowling alley. Bresler's son told me that not only had his dad witnessed the sinking of the *Jalisco* but had actually been one of a number of photographers sharing duty with Joe Rosenthal on the day Rosenthal snapped the iconic image of Ira Hayes and his fellow marines as they raised an American flag on Iwo Jima during World War II.

Houtz says that Bresler's dramatic photographs, presumably shot from *Polaris* from the backside of the waves, don't actually do the scene justice. On the *Jalisco*, the distance from the top of the bow to the waterline was thirty-two feet. Below the waterline lay another twenty feet of ship. Before the wave hit, it drew the water down probably ten feet below her waterline, thus leaving only ten feet of water for a cushion above the rock. "Kirkwood was hit by that wave, blown off the deck, then he might have fallen forty-five or fifty feet before he even hit the water," he said. "Then the waves landed on him."

The miserable men had a while together on the tug before being dropped off aboard the *Polaris*. Houtz had a broken rib, but O'Malley's internal injuries were worse, yet even he would quickly recover. Lesslie, Kirkwood, and Dan were merely, and incredibly, only battered and bruised. A small team of FBI agents reached *Polaris* by helicopter. They asked Houtz basic questions about his role in the operation, but reserved the bulk of their interrogation for Kirkwood, who faithfully spun his yarn about the ship striking the rock in the middle of the night and their then being forced to scuttle her. True or not, this is the story that stuck, and it kept Kirkwood and everyone else out of jail. Captain Cliff Miller and his crew, in fact, earned a Coast Guard citation for *Whitney Olsen*'s rescue work.

No mention of McMahan's five rock barges appears in Kirkwood's written account. Houtz maintained that when things went to hell, their captain wisely and quietly slunk back toward Ensenada. The seas, however, became so rough that at least one of the barges sank, carrying with it a fortune in boulders and a D-8 Caterpillar bulldozer, whose driver nearly drowned.

The Cortes Bank had almost killed Kirkwood, yet for months afterward, he lingered in the press, still broadcasting his grand plan for Abalonia. Other entrepreneurs floated their own nationalistic aspirations for Cortes Bank. Taluga would have been a glittering resort of three islands straddling the three shallowest shoals while the kingdom of Aphrodite would be built using classical Greek architecture with a government based on peace, tolerance, and love. Eventually, the U.S. Army Corps decided that the Bank lay on the U.S. outer continental shelf, and they forbade any further nation-building plans. Perhaps they realized that William Adger Moffett had actually claimed Cortes Bank for the United States back in 1911.

In the ensuing years, *Jalisco* would be pummeled by waves and eventually broken into three very jagged and very dangerous pieces. Despite this, she would become a popular spot for lobster and abalone divers. Today, she's still down there, biding her time as more and more of her sharp, rusty rebar becomes exposed by the ocean and scattered across Bishop Rock. At least a portion of her hull seems to rest in a spot that, as future arrivals would eventually come to realize, makes surfing exceedingly dangerous.

Joe Kirkwood remained baffled over the United States' demands that he cease and desist. "The most obvious reason for panic in the upper echelons of government is fear of another Cuba," he wrote. "While I can see their reasoning in this respect, the thought is almost laughable, for never was there a more fervent capitalist than I. And not only because I'm a businessman with modest financial success in a capitalistic system. I sincerely believe that the need for possession is inherent in each of us, and any system which denies man gratification of that need, must strip him of all incentive, and eventually, a reason to get up in the morning.

"Washington had nothing to fear from me. My thinking leaned rather toward a monarchy, and probably a constitutional monarchy, but in any event for the good of the people. One in which the spirit and intent of the law would be carried out, rather than, necessarily, the letter of the law."

It would be easy to dismiss Kirkwood's mania as crackpot, but he was not crazy. He eventually moved on, made a nice fortune for himself with a golf course on Kauai, and—as far as Houtz knows and I can figure—died of natural causes around a decade ago. Kirkwood was not alone in his obsession with Cortes Bank—for the freedom and riches it promised, for his "inherent" desire to possess and own it. This quality is shared by nearly every person who has encountered it. Friedrich Wilhelm Nietzsche described a similar, universal sort of need back in 1886 when he wrote, "Every superior human being will instinctively aspire after a secret citadel where he is *set free* from the crowd, the many, the majority."

You don't find much more freedom than by building your own citadel atop a sunken island.

Houtz completely understood and in his own way strongly shared Kirkwood's impulse, and it echoes in the stories told by Mel Fisher, Ilima Kalama, Bill Sharp, Sean Collins, and every surfer risking his thin neck on the Bank today. Kirkwood, in the grandness of his schemes, his efforts to defy if not destroy the wave, and his penchant to cast himself in the role of the benevolent monarch, perhaps fulfills the part of Ahab more than most, but as Melville makes clear, everyone on the *Pequod* hungered for the white whale.

And Houtz asks us to consider: Had he and Kirkwood reached Cortes Bank a day earlier, the seas would have been calm. With placid conditions, and time on their side, Houtz thinks it likely that he would have been able to work around the anchor issue by having *Whitney Olsen* simply motor *Jalisco* in the right spot and hold while ballast valves put *Jalisco* on the seafloor. With five barge loads of rocks then dumped off her bow, she would have been far better protected from the next day's waves. Houtz isn't certain she would have been protected *enough*. In fact, had he remained on board as planned, he probably would have died.

But Cortes Bank *would have become an island*. Fierce legal battles would have made global headlines. Most likely, if that happened, the only person to ever ride a wave out there might have been Harrison Ealey in 1961; Bruce McMahan's boulders would have surely ruined the break. Houtz, though, disagrees. He had planned on laying Abalonia's rocks in a long, slow, upward slope, and so, he said, "You might have ended up with the longest, biggest point break on Earth."

Today, Houtz doesn't talk much about what happened on Cortes Bank, even though it's one of the stranger episodes in U.S. maritime history with a cast of characters straight out of a movie. It's just not his way. Still, as he thumbs through his scrapbook, stopping at a photo of Joe perched out on the deck of the *Jalisco*, he shakes his head and tells me he relives the moment Kirkwood was blown off the deck every day of his life. Kirkwood's last words aboard the *Jalisco* still echo in his mind, their conviction so certain. "The wave's gonna go by me. It's gonna wash around me."

"I'm just sitting there looking out at Joe, just going, 'You're crazy,'" he said. "I still think the guy's crazy. But then again, so was I."

Chapter 5:

ROGUE
WAVES

The incidents in the life of a wave are many.
How long it will live, how far it will travel, to what
manner of end it will come are all determined,
in large measure, by the conditions it meets in its
progression across the face of the sea. For the
one essential quality of a wave is that it moves;
anything that retards or stops its motion dooms it
to dissolution and death.

—Rachel Carson, *The Sea Around Us,* 1951

As Rex Bank, James Houtz, and Joe Kirkwood discovered, if you find yourself above the Bishop Rock on a calm autumn or winter's day and don't heed the distant early-warning signs, the first breaking waves of a new swell might well be the last thing you ever see. Seemingly out of nowhere, they will rise up above you—deep blue and terrifying—like a line of tsunamis. Even if you get the motor on your boat started in time, you still might not outrun them.

In your last moments, two existential questions might or might not arise: How is it that a line of North Pacific skyscrapers have appeared seemingly out of nothing, and how damn high can those waves go? For answers, we can learn much from the experience of the USS *Ramapo.* In 1933, the crew of this navy oil ship was granted a sight previously seen only by God or the dead. Trapped in the midst of what was to the time the most powerful storm ever recorded in the North Pacific, the ship somehow survived being overtaken by what was believed to be a physical impossibility: a wave 112 feet high. This remains the biggest wind-driven swell ever reliably measured by an eyewitness from a position on a ship in the open ocean. When this same swell eventually reached Cortes Bank, it created waves that were in all likelihood taller, and today the Bank is considered almost uniquely without an upper limit. The Bank not only produces the largest surfable wave on the planet, but no one really knows just how high a breaking wave, under the right conditions, might reach—a thousand feet has been tossed out as possible.

The storm that created the *Ramapo*'s record-setting wave had its genesis in the loneliest reaches of the North Pacific—a vast, malevolent swath of ocean below the Aleutian Islands that lies directly in the track of a wintertime jet stream whose high-altitude winds can exceed two hundred miles per hour. Reports from particularly violent tempests in this zone are scarce—captains dodge them or don't survive to tell the tale. That's why the experience of the *Ramapo*, recounted by its executive officer Ross Palmer Whitemarsh in a 1934 article titled "Great Sea Waves," is singularly unique. Understand the gauntlet run by these sailors, and you can begin to comprehend how a pair of equally mammoth storms seventy-five years later could send incomprehensible giants roaring into Waimea Bay; Maverick's; Todos Santos, Mexico; and of course, the Cortes Bank. While mariners have learned to chart the conflux of weather, swell, and current to avoid these beasts, surfers have schooled themselves in these arcane arts for the exact opposite reason.

Ross Palmer Whitemarsh was born in Olympia, Washington, in 1895. He was a mathematical genius who graduated from the U.S. Naval Academy at Annapolis in 1918 and went on to lead the sort of life that is the fodder for a Tom Hanks epic. Yet in "Great Sea Waves," Whitemarsh recounts his almost unbelievable story in a somewhat dry, scientific manner. In short, the account doesn't reveal much about *who* he was or what he was thinking during the experience. Fortunately, a few of his close family members were happy to shed light on their legendary patriarch.

His grandnephew James Whitemarsh—today, an auto shipping executive in his sixties who lives in West Palm Beach, Florida—remembers idolizing his great uncle before they even met. "I was reading this story of men and survival in the sea in a Time/Life Book," James said during our interview. "And I was shocked to see his name. When I asked my mom who he was, she said, 'Oh, that's your uncle Ross.'"

James met his great uncle shortly thereafter in 1957, when James was eleven. By then a rear admiral, Ross Palmer had just retired and was visiting the Washington State side of his family. He was fit and robust, but James had expected someone taller.

"He was short, perhaps no better than five foot seven," he said. "But being an officer and being so short, he had to have a big personality."

The admiral regaled his young nephew with incredible stories of survival. At twenty-five, Ross Palmer had been assigned an unlikely duty as the senior naval officer on board the *Dwinsk*, a British freighter that ferried American troops to the European front during World War I. *Dwinsk* was returning home on June 18, 1918, when a German U-boat sank her around six hundred miles off the coast of Norfolk, Virginia. Everyone escaped to lifeboats, but rather than letting the men sail off, the U-boat captain decided to toy with them, using their lifeboats as decoys to attract other ships to torpedo—as a sniper might lure enemy soldiers by shooting one of their fellows in the knee. After a few harrowing days—during which other Allied captains learned what was afoot and refused to approach the lifeboats—the German captain left the men to die. For the next eleven days, through torturous drought and a journey into the eye of a hurricane, Whitemarsh was the glue that bound twenty desperate men.

"They realized that the sail on their boat was rotten," said James. "But just as the men were getting ready to jettison the sails, my uncle told them to save them to catch the rainwater. In the end, that's what saved them."

By the time Ross Palmer Whitemarsh set sail aboard the *Ramapo* fifteen years later, he was a seasoned mariner, a husband, and father of two young girls. One of his daughters, Francis "Taffy" Wells—today, a delightful, plucky octogenarian living in Honolulu—helped me reconstruct her father's most legendary adventure.

Ms. Wells grew up both adoring and fearing her devoted father, whom she recalls as a fanatical golfer and strict taskmaster. "His favorite saying was 'order, counterorder, disorder,'" she said. "I heard it hundreds of times. If you didn't follow orders, there'd be disorder, and he *hated* that. I remember shagging golf balls for him as a little kid. I'd sit off to the side behind a palm tree, and when he'd hit all the balls, I'd run out and pick 'em up. And you know, he had this dry, British humor. He'd tell a joke with a completely straight face. You'd sit and think about it for an hour, and then you'd just start laughing like crazy."

Ms. Wells was three when her father set out from Manila aboard the *Ramapo*. The ship was a Patoka class oiler, built in 1919 at Newport News Shipbuilding and Drydock. She was 478 feet long (about one and a half football fields), sixty feet wide, and weighed 17,000 tons (better than three times heavier than *Jalisco*). She was also low slung and stable, drawing twenty-nine feet of water. Despite a pair of twin 2,800-horsepower steam turbines, she was, like *Jalisco*, a plodding ship. Laden with seventy thousand barrels of oil, she was barely capable of ten knots. This made her two-thirds as fast as the USS *Constitution* or, say, the *Pequod* under full sail.

Between 1929 and 1934, *Ramapo* made a great many crossings from San Diego to Manila. According to Ms. Wells, the journeys became incredibly monotonous. "The crewmen kept getting into fisticuffs and trouble," she said. "Dad was trying to figure out something that would involve the whole crew and get them to stop bickering."

Whitemarsh and *Ramapo* Captain Claude Banks Mayo had an advanced echo sounder brought on board, and they decided to teach the crewmen to use it and interpret the results. On each trek, the ship took a slightly different route, and after 17,239 soundings, her crew eventually produced a huge 3-D plaster of paris map of the midlatitude Pacific seafloor. It was stunning in detail, unveiling two trenches, the Nero and Ramapo Deep (now called the Japan Trench) that were more than thirty thousand feet deep. Mayo wrote that the map revealed "a submerged continent, with mountains, river courses, and plateaus, at an average depth of one mile stretching from the Hawaiian to the Barin Islands, east of the coast of Japan." The map and its later, more-refined iterations became an instrument not only for geologists and oceanographers but for early surf forecasters like Walter Munk, who wanted to know how waves interacted with the seafloor. Countless mystics came to believe that Whitemarsh and Mayo's map in fact exposed the outline of the lost continent of Lemuria.

When Whitemarsh set out in late January of 1933, the longstanding dogma since the days of Archibald MacRae was still held as a scientific truth. Wind-driven waves could only grow as high as 60 feet. Anything higher would be pulled down by simple gravity. Thus, all the tales of taller waves were categorized as wild-eyed, rum-laced myth.

After delivering a full load of fuel to the Pacific fleet in Manila, the *Ramapo* was following another of her great circle sounding routes. She was well north of the Hawaiian Islands and their ports, and her sole link to the outside world was a shortwave radio, whose antenna was stretched between a pair of masts.

Had that antenna given Whitemarsh the benefit of satellite imagery, he might have asked Mayo to steer a different course, for the *Ramapo* was about to become a tiny pawn in a global atmospheric upheaval. A La Niña weather event of epic proportions was already underway. Cold and snowfall records were being set from Belfast to Kamchatka to Manhattan. On January 11, a hurricane-force gale had lashed the entire California coastline, destroying 130 oil derricks and spawning nearshore waves that swept sailors from the decks of four U.S. warships—including an aircraft carrier. Farther north, blizzards gave Crater Lake, Oregon, a January snowfall of 256 inches, and the mountain town of Seneca, Oregon, recorded temperatures of forty below—records that still stand. On the upside, a newly sworn-in President Roosevelt was about to repeal prohibition, allowing Americans to legally drown their sorrows, and they'd need to. That summer, twenty-one tropical systems—a record lasting until 2005—would form in the Atlantic Ocean, devastating the Chesapeake Bay, North Carolina, and Texas. By November 1933, ceaseless winds would begin scouring the topsoil from drought-stricken farms in the Dakotas and dropping red snow on Chicago. It was the dawn of the Dust Bowl.

The January storm began off the eastern coast of Russia, when a vicious surge of polar high pressure blasted Siberia and swept out across the Pacific at fifty miles an hour. In the wake of this surge, the temperature in a barren outpost called Oymyakon plunged to ninety degrees below zero—the coldest temperature ever measured in the Northern Hemisphere.

Within a few feet of reaching open Russian waters, minute vertical pressure changes in the air caused tiny, almost invisible deformations—scientists call them capillary waves—to form on the water's surface. These gave a rough texture for the horizontal winds to grab, causing the water molecules to begin to vibrate in a circular motion. By a hundred yards offshore, that circular motion had manifested in the form of diminutive wind waves, whose peaks formed rows of miniature sails and whose troughs carried a rotating eddy of air that furthered them along.

By the time these sails, or swells, had been pushed twenty-five miles offshore, the insistent wind had morphed them into orbital columns—somewhat akin to a line of logs rolling downhill—around 11 feet high, with a period between their troughs and crests of seven seconds. Most of the energy of these swells was not at the surface of the water, but below it, reaching down at least as far as forty feet. After another 250 miles of wind, the swells were more gently rounded but were now averaging around 27 feet high with a far longer thirteen-second period and an energy column *four hundred* feet deep.

And they kept growing.

The jet stream pushed the Siberian front as far as the International Dateline a few days later. There it began to interact with one storm centered just above Hawaii and another near Dutch Harbor, Alaska. The dense air of the frigid high pulled a vast plume of sultry, low-pressure air from a dying tropical depression far into the North Pacific. The winds around the high funneled into the lows, fueling a precipitous drop in barometric pressure, a condition meteorologists call "explosive cyclogenesis," or just as often "a bomb." That bomb detonated above the *Ramapo* on the morning of February 6. Winds went from whining to screaming through her rigging, and the sea surface became streaked and marbled with wind-driven foam and spray. "The elements began to give evidence that something out of the ordinary was about to happen," Whitemarsh noted dryly when writing about it later.

In the space of a few hours, the waves rapidly grew from simply large to mountainous juggernauts fifty feet high, a quarter mile wide, reaching *twelve hundred* feet deep under the surface. Their forward speed was, somehow, actually slightly faster than that of the sixty-mile-an-hour wind.

When the lows consolidated into a single monster system, *Ramapo*'s barometer plummeted to 28.40—a record for a North Pacific winter storm—

and sustained winds reached 70 miles per hour. Since hurricanes can have sustained winds of 150 miles per hour, this might not sound very dire. However, a hurricane's core wind field is usually no more than a hundred miles across, while winds over 110 miles per hour actually blow the tops off waves and thus, in effect, reduce their height and power. This cyclone was at least as wide as North America, and the hurricane-force winds spanned half of the North Pacific. In the deep troughs of swells, the air was eerily calm, but when *Ramapo* was born to the crests every minute or so, she was assaulted by the gale. Apocalyptic squalls occasionally gave way to moonlit skies and awestruck viewing.

To survive, the *Ramapo* had to maintain the same direction as the swells, keeping the waves at her stern and running like hell. Yet compared to waves moving at six times her top speed, she was almost standing still. "The vessel was dwarfed in comparison to the seas," Whitemarsh wrote. "It would have been disastrous to have steamed on any other course."

As the seas grew, the *Ramapo*'s single screw propeller would lift clear of the water each time it crested a swell, causing the engines to rev dangerously, and Captain Mayo was forced to reduce power. This was perilous. Were the ship not being driven forward with sufficient force, she might wallow sideways across the westerly winds and waves. Had this happened, she would have been driven nose-first into the seething Pacific and disappeared within moments. Instead, she was quite literally surfing for her life on the downslope of each passing swell.

James Whitemarsh recalled, "He told me they were simply running from the waves. In all his years at sea, he never saw anything like them again."

Serendipitously, the *Ramapo* turned out to be the perfect size and dimension to survive. She was 478 feet long—two Boeing 747s placed nose to nose— and this length fit snugly between the waves. Were she, say, 880 feet, the length of the *Titanic* or a modern-day container ship, the situation might have been terminal. Atop the crest of a 1,200-foot-long wave, her bow and stern would have lost buoyancy and drooped down—a condition called *hogging*. Fifteen seconds later, the situation would have reversed, with bow and stern supported, but her amidships bearing tremendous weight—a condition called *sagging*. These forces can easily result in a catastrophic set of side to side rolls; if they don't, they can essentially snap a ship in two.

Instead, as Whitemarsh later noted, the conditions actually made for ideal observation. Swell and wind approached from exactly the same angle. *Ramapo* was thus not faced with the specter of wind-driven waves approaching from sideways angles—a nightmarish condition that can cause waves to ramp up into treacherous peaks. She was steaming across a vast, undulating ribbon of seemingly impossible energy.

For the most part, Whitemarsh leaves us to wonder what he and his ninety crewmen thought and did while they endured conditions straight from a Jules Verne novel. Surely, he and Captain Mayo would have filled the *Ramapo*'s ballast tanks, so she would ride low and stable in the water; general quarters would have been sounded, with a contingent of men ready to cut loose the lifeboats without a moment's hesitation (as if there was a hope in hell in a lifeboat). And yet beyond maintaining the ship's speed and position and their readiness to evacuate, they were mostly along for the ride, free to gape and wonder and pray. Over and over, hour after hour, the ship struggled up one hill and into the teeth of an unspeakable gale, and then, every fifteen seconds, she slid down the next into a dead calm, shouldered between racing mountains.

As earlier scientists theorized, there comes a moment during the evolution of a storm when gravity asserts its dominance over a swell. It is simply carrying so much water that its weight prevents it from growing any higher. At this point, the sea state is said to be *fully developed*. In modern theory (that is, what is generally accepted today), the maximum height for a fully developed, single open-ocean swell is 78 feet high with a period of twenty-three seconds and a velocity of seventy-seven miles an hour. This occurs after a seventy-knot wind has been pushing on the swell for a thousand or so miles.

However, the *Ramapo* was in just this situation, and it would soon record a single rogue wave nearly 50 percent higher still. How can this be?

Simply put, this tidy calculation of potential wave height is insufficient to account for the almost incomprehensibly complex physics of wave generation, and science has come to accept that what appears to be a *single* wave is, in some cases, anything but. So while open-ocean waves *should* fit this formula, exceptions occur all the time and for a whole host of reasons. Not only do waves interact with the swirling atmosphere and respond rapidly to changes in wind speed, but because waves of different periods run at different speeds, they regularly run through and over their brethren. One wave might cancel out or highly amplify the energy of another, a process scientists call nonlinear transfer. A 30-foot, 10-second wave 100 feet deep might be overtaken by a 60-foot marauder 20 seconds long and 1,200 feet deep; this might result for a short time in a single 90-footer we call a "rogue wave." Or a line of big swells might crash headlong into a strong ocean current like the Gulf Stream—and stack up like a chain-reaction crash on a freeway into a series of steep, deadly peaks. The result is a temporary but highly unstable wave, actually a trio of rogues, that mariners call "The Three Sisters." These climb far higher—perhaps even two to three times higher than the surrounding swells—and their troughs can seem so fathomless that they have been called "holes in the ocean." When one wave runs over the top of another, it can actually rise up and shoal across the slower wave's back, producing a gigantic breaker that, like the Cortes Bank, appears in the empty open ocean. These waves are not uncommon at all. A famous 2001 European Union study called "Max Wave" used a network of global satellites to determine that, at any single point in time, there are probably eight or nine such rogue waves—either in single or "three sisters" form—coursing through the world's oceans.

In his article, Whitemarsh describes several similar encounters then known to mariners of his day. In 1837, Jules Dumont d'Urville claimed to have seen a 100-foot wave while sailing around the Cape of Good Hope. In 1861, a 100-foot-high lighthouse off England's Isles of Scilly was struck by a wave that tore off the bell in her tower. The Tillamook Light in Oregon had seen days when waves blasted rocks through windows 133 feet high.

Even Christopher Columbus told of giant, seemingly impossible waves during his third trip to the New World in 1498. Columbus was leading a fleet of six ships through a stormy, narrow passage at the southern tip of Trinidad. He heard "a fearsome roaring" and turned to see a wave higher than his over-60-foot masts bearing down on the flotilla. It lifted all the ships higher than anything the admiral had ever seen and then dropped them into a frightening trough, burying them with foam and spray. He named the passageway "The Mouth of the Serpent."

The passengers aboard the *Annie Jane* were not so lucky. On September 28, 1853, the steamer left Liverpool for Canada carrying more than five hundred emigrants to a new life in America. But as *Annie Jane* was passing the Hebrides islands in the dead of the night, she was struck by a frigid, black wall of water that came out of nowhere, collapsing her poop deck and instantly crushing two hundred people.

In May of the same year, the fifty-two-foot yacht *Mignonette* set sail from Southampton, England, for Sydney, Australia. But off Africa, a hurricane-force gale pushed great northerly swells headlong against the south-flowing Agulhas current, which sweeps past the Cape of Good Hope. Amid a pileup of swells, the struggling *Mignonette* finally succumbed to a single, colossal peak again higher than her masts. Nineteen days later, starving and adrift aboard a leaky dinghy, the desperate captain and two crewmen decapitated and devoured their dying seventeen-year-old cabin boy. The trio was miraculously rescued four days later and carried back to England to be tried for murder and cannibalism. Following one of the most macabre court cases in maritime history, one man was acquitted while two spent a mere six months in prison. But the sensational case forever outlawed a gruesome, if occasionally necessary, "Custom of the Sea."

Whitemarsh doesn't give heights of the waves that regularly swept beneath the ship but instead noted that they typically moved at sixty to sixty-six knots. "Probably no two seas [waves] were identical in length and height," he wrote. "They varied from 500 to 750 feet in length of sides, or total wave length of 1,000 to 1,500 feet, as measured by the ship itself and the seaman's eye. This is verified by motion picture film taken during the morning watch. The *Ramapo* is 477 feet ten inches in length. For purposes of illustration, a conservative wave length of 1,180 feet is assumed. It was noted that the ship's entire length glided down the lee slope of waves an appreciable time before the next crest overtook the stern." [Unfortunately, the film footage Whitemarsh refers to seems to have disappeared into a black hole at the U.S. Naval or National Archives.]

After twenty-four hours of racing up and down over behemoth waves, Whitemarsh posted Lieutenant Frederick C. Marggraff to watch and measure the swells from an ideal spot atop the pilot's house. At around 3:30 A.M., Marggraff stood dumbfounded as a billowy monolith eclipsed the moon. A great wave had collected what seemed the entire ocean in its maw. It stacked up behind *Ramapo*, nearly a quarter mile long from its trough to its crest, and moved much faster than the other swells. As it overtook the ship, the *Ramapo* began to slide forward like a surfboard descending a wave. Whitemarsh and the crew held on as *Ramapo* pitched forward a harrowing twenty-four degrees. Fortunately, the great swell was not breaking in a top-to-bottom fashion; it was simply so large that a portion of its upper reaches submerged the ship's stern and cascaded down her deck. Had the swell in fact been breaking over the backs of its fellow waves, *Ramapo* would have been completely buried and almost certainly sunk. Instead, she rocketed forward as if atop a sea of ball bearings. In the canyonlike trough, her bow dug in but not so deep that she was pitchpoled forward. Despite the tumult, Lieutenant Marggraff somehow kept his nerve. The five-foot-eleven officer set a line of sight between the pilot's house and the crow's nest. Using basic trigonometry, he calculated a true wave height of 112 feet, or 34 meters—that is, a wave roughly ten stories high. This remains the largest wave ever observed by a human being from the deck of a ship, and the first reliable documentation of a wave greater than 100 feet high.

Since that time, there have been numerous terrifying and reliable measurements of waves approaching, but none officially "besting," the *Ramapo* wave's height when measured from an actual sea-level position. In 1995, the North Sea oil platform *Draupner* had a mammoth rogue wave 95 feet high sweep beneath her from seas that were averaging only around 30 feet. Nine months and

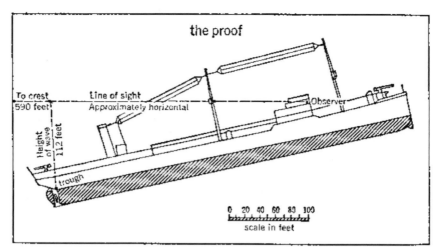

the proof

To crest
590 feet

Line of sight
Approximately horizontal

Observer

Height of wave 112 feet

trough

0 20 40 60 80 100
scale in feet

Fig. 37–1.

eleven days later, Captain Ronald Warwick was captaining the *Queen Elizabeth II* through 40-foot seas off the coast of Newfoundland around the periphery of Hurricane Luis. At 4:10 A.M., a Canadian NOMAD buoy nearby bobbed over a wave 98 feet high. Moments later, Warwick gasped. An eerie line of white phosphorescence loomed across the horizon directly ahead. It appeared, he famously wrote, "that the ship was heading straight for the white cliffs of Dover."

The ship plowed into a wave that smashed seventy feet above her waterline. She then careened downward into a breathtaking drop on the wave's backside before facing an even more nightmarish peak. This wave exploded through thick windows of her main lounge, two levels above her eighty-foot-high main deck. Had the ship been sidelong to these waves, she would have surely rolled right over.

The biggest weather-driven wave *ever* measured seems to have been generated during a disastrous gale whose 70- to 80-foot waves sank five sailboats competing in 1998's Sydney to Hobart race. An Australian helicopter went in to rescue the crewman of the stricken vessel *Kingurra*. As the bird hovered in position near the boat, pilot Darryl Jones saw a massive wall of water bearing down and made an emergency ascent to 150 feet. When the wave passed beneath him, his precise radio altimeter measured a mere ten feet of elevation. The wave was 140 feet high.

Miraculously, *Ramapo* was drawn backward up the face of her mammoth wave and eventually over its crest and down its backside. The wave passed on, heading directly for the Cortes Bank, carrying an almost incomprehensible amount of energy in its belly, the equivalent of 29.6 billion watts. As a comparison, on a hot summer's day, the entire city of New York consumes only about 11 billion watts.

Ross Palmer Whitemarsh survived another forty-four years. On December 7, 1941, he was in charge of a minesweeping division. From the deck, of the USS *Gamble*, he watched bombs drop around and narrowly miss *Ramapo* as she lay exposed in Pearl Harbor. For all his life, he swore that one of the boats under his command sunk a torpedo-laden miniature Japanese submarine in the hours before the attack. Such a damning revelation would cast a harsh glare on American commanders who claimed the attack was a complete surprise. "But the Pentagon and whoever wouldn't believe him because he had no proof," said his daughter, "Taffy" Wells. "Then a few years ago, they found the sub."

Whitemarsh went on to lead mine-clearing operations in Iwo Jima and Okinawa, while facing intense artillery barrages and a kamikaze attack that nearly sunk his ship. His service earned two Legions of Merit and a slew of other citations.

Not long before Whitemarsh died in 1977, his grandnephew James visited him one last time. Ross Palmer was still mowing his own yard. James said, "He was a cool customer. He never once conveyed to me that he feared for his life."

I asked Ms. Wells if her dad ever said he was frightened during his time aboard *Ramapo*, or later during the war. "Dad went over huge waves," Ms. Wells said, "had ships blown out from beneath him. But no, he wasn't scared. My dad was never scared of anything."

The descriptions of Whitemarsh by his relatives parallels something I would come to recognize in people from Flippy Hoffman, Jim Houtz, and Harrison Ealey to the big wave surfers facing ten-story rogues atop the Cortes Bank today. In short, even if they do feel fear, they are somehow able to overcome it—to become almost analytical in the face of life-threatening conditions that would leave most of us paralyzed.

At the end of "Great Sea Waves," Whitemarsh wrote: "Since time immemorial, seafaring men have been telling the world in their inarticulate way that storm waves attain heights which seem incredible to the rest of mankind. The privilege of viewing great storm waves of extreme height is a rare one indeed. Furthermore, we have no assurance that the highest waves of the ocean have been observed or measured. If such a wave should ever be encountered, it is probable that all hands would be chiefly concerned with the safety of the ship to the exclusion of any scientific measurement of the phenomenon.

"Perhaps authorities in the past have been radically conservative in the treatment of sea waves. A 60-foot wave as the highest of all time lacks conviction. . . . The theory and law of waves are excellent guides, but, inaccordance with the present custom, if the laws cannot be enforced, they should be repealed."

By February 9, after about three days of doing nothing but running with the waves, and after a week of being enveloped in the gale, the *Ramapo* began to make a slow, steady exit from the storm. The storm was tracking a bit more to the north, and *Ramapo* managed to steam on a more southerly course that would soon take her right past the Cortes Bank and on to San Diego. Meanwhile, the monster swells whipsawed out far ahead of her, spreading out like ripples from a rock thrown on a pond, which dropped in height, but not in depth. The stronger, longer swells passed through the shorter, weaker ones and raced to the east. They would utilize a strange, almost evolutionary logic to self-organize during the next fifteen hundred miles by period, forming sets or trains that held anywhere from five to fifteen waves. A train of such waves is a bizarre exercise in physics. The fastest train—waves with periods that sometimes grow to better than twenty-five seconds—rolls toward shore at around thirty-five miles an hour, yet the *individual* waves within that train travel at twice that speed. They continually roll from the front of the train to back—like the tread chains on a tank.

After a couple of days, this new swell formed a very well-defined front—a bulge several thousand miles across—and the most powerful waves peppered its leading edge in mile-wide ribbons.

It's rare that a person sees the leading edge of a deep swell like this. If you're aboard a ship in calm weather and deep water, you might become aware first of an increase in very long, seemingly slow rollers, followed in an hour or so by the appearance of higher, somewhat steeper waves that seem to stretch across the horizon. This is what James Houtz noticed but failed to identify quickly enough aboard the *Jalisco* in 1966.

The vast majority of coastal landmasses don't see these forerunner waves. They run so deep that they scrape across the continental shelf and are refracted or deflected away from land, losing much of their energy in the process. But there are places, like Cortes Bank, where the seafloor transitions rapidly upward from abyssal depths. Forerunners are drawn to such spots like rays of sun through a magnifying glass. They initially stir the water in low half-minute or longer undulations, but as their periods shorten, tremendous breakers can appear almost completely without warning.

It's interesting to note that, due to simple meteorology, the prime, raw energy of the strongest North Pacific storms—open-ocean swell energy between 50 and even 100 feet—typically doesn't travel much below the latitudes of Washington and Oregon. Thus, the *Ramapo*'s giant wave likely dropped to a solid 25 to maybe 35 feet well before it careened headlong onto the Cortes Bank on February 11, 1933. The Bank's two unique characteristics—that a wave of almost limitless size can break here, and that the Bank, as has been proven, magnifies long period swells into breaking waves between four and five times a swell's height—means that were the Cortes Bank situated a hundred miles off the coast of Seattle, it might regularly spawn waves *at least* 200 feet high.

Instead, the *Ramapo*'s wave probably reached a breaking height somewhere between merely 100 and 150 feet over the Cortes Bank. But like a giant tree falling in a forest, its thunderous fury was released in essentially undocumented privacy. This was understandable. Even if someone had known it was coming, in 1933, no veteran fisherman, no pioneer surfer—no human in his or her right mind—would have set out to confront such a deadly freak of nature.

But nearly six decades later all that would change.

Chapter 6

MAKING
THE
CALL

As he stood hovering over you half suspended in air,
so wildly and eagerly peering towards the horizon, you
would have thought him some prophet or seer beholding
the shadows of Fate, and by those wild cries
announcing their coming.

'There she blows! there! there! there! she blows! she blows!'

—from Herman Melville's *Moby-Dick*, 1851

By 1985, when Cortes Bank first blinked onto his big wave radar, Larry "Flame" Moore was just as obsessed with chasing down the best swells nearer to shore. Nor were Bill Sharp, Sam George, and George Hulse his only comrades in arms. Flame had enlisted a group of reliable, hungry California chargers into a loose confederacy that friends and rival photographers only half-jokingly called "Larry's Army." The recruits came to include a cadre of hot Orange County surfers—guys like Chris Mauro, Dave Parmenter, and Terrence, Joe, Brian, and Pat McNulty, as well as a fastidious and fanatical redheaded shredder named Mike Parsons.

Flame would assemble his troops and photograph them dissecting the glistening, front-lit waves at Salt Creek in Dana Point, the point break peelers at San Clemente's famed Trestles, or perhaps they'd make the two-and-a-half-hour drive below the U.S.-Mexican border to charge the heaving barrels of Baja Malibu. Flame was dictatorial, demanding that his subjects hold themselves to professional standards—an idea then completely at odds with the popularized image of the Jeff Spicoli slacker/stoner immortalized in the film *Fast Times at Ridgemont High*. If you weren't on time, you were left behind. If your board or wetsuit didn't have vibrant color, forget it. And it didn't matter if you were Laird Hamilton, if you disappeared into a smoke-filled VW bus when you were supposed to be out surfing, you were going to catch pure hell.

Still, Flame demanded nothing of his subjects that he didn't demand of himself. He considered it requisite to pull on his wetsuit and swim fins and swim or paddle into the heaviest conditions on a boogie board clutching a heavy water housing for his camera. The only way to get stellar barrel shots of surfers was to put yourself right in the teeth of a blue-green cyclone and prepare for the detonation.

Oftentimes, an eager understudy of Flame's named Robert Brown would also show up and snap differently angled shots of the same waves—particularly at Salt Creek. Flame would bark at Brown for poaching *his* shots, yet Brown says it was mostly bluster, aimed at determining whether Brown could take the heat. Flame was an egalitarian dictator who saw talent and drive in Brown and was just as likely to use Brown's shots as his own in the magazine. At the *Surfing* offices, Flame would offer encouraging critiques of Brown's work and that of a great many upstart lensmen, offering advice on what lighting worked—most desirable was a front-lit condition everyone came to call "Larry Light"—and where surfer, ocean, sponsor logos, and points of land needed to be for an advertisement, a double-page spread, or the hallowed cover. The tech-savvy Flame also recognized great value in the fact that Brown possessed a boat for offshore expeditions. "We had a funny relationship," Brown says. "I was working under Larry, but I was also his competitor. He was doing all he could to sabotage me."

Brown had grown up surfing the sandbar barrels at Salt Creek. Early on, he particularly hated Larry's top model, Mike Parsons. It wasn't that Parsons was a jerk. He was a damn anachronism—polite to a fault and infuriatingly skillful. "We'd surf Gravels and he'd come down from Laguna with Chris Mauro and maybe George Hulse. He didn't fit the profile of a pro. He was this skinny red-headed stepchild with freckles. But he had his *perfect* wetsuit with his sponsor's logo airbrushed on it, and he pulled these *perfect* off the lips. We'd try to vibe him, but he out-surfed us so bad that it just didn't matter."

Parsons and Flame clicked particularly well. Not only was Parsons attentive and scrupulous, he was equally driven and obsessively crazed for the next surf fix. He was ascending the competitive ranks and was fearless when it got big. In Flame, he found the perfect documentarian—and the perfect friend.

Flame's early tenure at *Surfing* also coincided with the first forays into surf forecasting of a stocky, ruddy-faced photographer named Sean Collins. Long before he would become surfing's one true media mogul, Collins was a brilliant high school dropout with an addiction to waves. He thought nothing of dropping everything—bartending or table-waiting jobs, girlfriends, responsibilities—to travel to the dustiest, remotest beaches of Baja California. He'd surf his brains out and take photos to make a few dollars on his return home. Collins lived feral for long stretches through the late seventies and early eighties, sharing barren campsites with Sam George and a few other friends. This was the sort of lifestyle that Harrison Ealey, Rex Bank, and Ilima Kalama treasured in the fifties and sixties, but in no small measure due to Collins, it was about to become permanently endangered.

Unlike most surfers, Collins wasn't content to simply wait for waves. He developed a mania to understand *where they came from* and, just as important, *when they would arrive.*

Still today, Collins speaks in a soothing, Southern California surfer's tone that betrays none of his intensity. He readily admits that his early motivations had nothing to do with making a living but, he says, were based on straight up fascination with swells and pure selfishness. He wanted to score waves—particularly summer swells that originated off Antarctica. Basically, he wanted to beat everyone to the punch.

Collins had become intrigued years earlier by the groundbreaking work of Dr. Rick Grigg, a Hawaiian big wave surfer and navy forecaster, and a navy scientist named Walter Munk.

In addition to figuring out why the moon only showed one side to the earth, Walter Munk used his mathematical genius to develop a measurement scale for wave energy based on height and period. He coupled swell measurements taken by the pilots of Pan American World Airways "Clipper Ship" seaplanes, which crossed the South Pacific in the 1930s and 1940s, with his own hard-fought understanding of the physics of wave propagation and decay (which refers to a swell's loss of energy as it radiates out over long distances). He compared these PanAm records with weather maps of storms in the distant latitudes above Antarctica, a perennially tempestuous zone known as the "Roaring Forties." By jibing the two, he figured out how a swell radiated across the ocean. This resulted in critical, life-saving forecasts for Allied World War II landings in North Africa and Normandy.

In the mid-1950s, Munk helped marine landing parties understand waves along the beaches off Camp Pendleton, south of San Clemente. He then traveled sixty miles offshore to San Clemente Island, where he deployed deep-water pressure sensors that measured a swell's power and direction. Among his less-celebrated findings was the fact that from certain angles of approach, San Clemente's swells were far smaller than you would expect. Munk correctly postulated that their energy was being "shadowed" or blocked by a pair of big damn obstacles—namely, the Cortes and Tanner Banks. More celebrated and revolutionary was the discovery that some waves that reached California weren't actually generated in the Pacific at all, but west of Australia in the Indian Ocean. The swells made a great circle around Australia and New Zealand and literally came ashore from halfway around the world.

Sean Collins saw a documentary film about Munk's work and it set him on a path.

"Munk scientifically confirmed to us surfers what we already knew—that some of our best swells were from the Southern Hemisphere. But back then for us it was all coconut telegraph—someone talking about a big storm in New Zealand and trying to follow it. Then getting reports from Tahiti, with the swell seven days from us, then Hawaii with it two days away, and then anticipating the swell's arrival in California. But it was really hit or miss. Not all those swells would even make it to California.

"Back in the seventies nobody really had a clue about real forecasting. I couldn't use Munk's work since I wasn't a calculus whiz—I just wasn't smart enough. And most of the oceanographic papers that I found in the Federal Depository libraries and weather service offices were far too complicated for me to fully understand. So Southern Hemisphere swells were still just mystical rumors. We're relying on storms sending waves from between five and ten thousand miles away. All of a sudden a swell would just show up. It would be flat and 1 foot one day, and the very next morning it would be 8 to 10 feet. It was like Christmas. That's where the whole 'surf's up' thing came from. You drop what you're doing, run to the beach, and jump on it—because it was a 'good today, gone tomorrow' kind of thing."

One bright day in late 1979, Collins posted up on the roof of his house and flipped open a notebook to watch the waves running up against the Seal Beach jetty. His work quickly evolved into a fanatical obsession—a daily log of height, angle of approach, seconds between individual waves, minutes between sets of waves, and the number of waves in each set. Collins then bought a clunky old marine fax machine capable of decoding the crackling beeps and chirps broadcast via shortwave from Christchurch, New Zealand. The clearest signal arrived at three in the morning, and the printouts of swirls and isobars gave only the crudest information about wind speeds and pressure gradients. But the images could be held up one after the other, allowing for a rough animation of how storms and their winds were evolving. He learned, like Munk, to hindcast.

"I was kind of like my own virtual buoy," he says. "And that just really taught me about swells. Southern Hemi swells were the hardest to forecast. They were like a black hole. You could hardly get satellite photos back then. I'd get those weather-fax charts and keep a library every single day. You could backtrack to determine size, timing, swell period. When we'd get really good waves, I could measure the swell period and tell how fast the swell actually traveled in deep water, so I could reverse it and go back to the point of origin. That would tell the location and date where those waves came from by the speed of their travel. I taught myself what to look for—this storm didn't look all that good, but the swell it produced was incredible—why did that happen? Over a few years, I got to a point where I was about 75 to 80 percent accurate."

Collins amazed his friends when he showed up unexpectedly one day at their Baja campsite with his weather-fax. He strung the antenna wire around his tent and juiced the machine with his car battery. He told Sam George and his fellow campers, "Watch this, tomorrow it's going to be six to eight feet."

George says, laughing: "We were living in a tent, and he was getting faxes from the weather service. It was just unbelievable technology. All I knew before Sean was that on the great south swell of '75, we were like Indians. When the moon came full and the day grew long, the great waves would come rolling in from the south. I didn't even know what the hell a swell was."

"Those were fantastic days," says Collins. "Nobody else knew anything, and I'd just get tons of waves. It was like voodoo."

"You know, in eighteenth-century Hawaii, there was this whole cult of surf priests, or kahunas," says George. "They had a temple on Waikiki Beach. By monitoring natural things like water color and temperatures or bird flights indicating a faraway storm, they could predict the surf. The fantastic thing was, when the waves were going to hit, the kahunas would send kites above Diamond Head to tell the villagers, 'Hey, the surf's up!' The villages would empty as the people rushed to Waikiki. Sean became our kahuna."

Flame soon recognized that Collins's predictions were becoming accurate enough to bank on and offered him a five-hundred-dollar-a-month forecasting retainer. Collins would give the heads up, and Flame would position photographers and surfers along the coast.

The mid- to late eighties were heady days for *Surfing*. Flame was getting baffling and frustrating drops on his archrivals at *Surfer*, whose offices were just up the road in San Juan Capistrano, and Flame's two young editors were talented provocateurs. On the one side was Sam George—an earthy, long-haired, and loudly opinionated soul surfer from the Bay Area. On the other was Bill Sharp—a brash and equally opinionated neon-clad ripper from Newport Beach with his halo of spiky bleach-blond hair. Their night-and-day lifestyles provided a yin-and-yang balance for a magazine targeting young readers. "Bill dyed his hair and wore tiger-striped spats," laughs George. "He was raging at the clubs and ska dancing to the Specials. I was listening to Loggins and Messina."

If George tended to be self-serious and even self-important in his pronouncements about the surfing life, he was also generous, gregarious, and always open to other opinions. And in Bill Sharp, he found the perfect philosophical and rhetorical foil.

"When Bill started full-time at the mag, I had a little trepidation because we were so different," George says. "But we saw right away that he had a really wry sense of humor, and over that, we established a bond. We didn't butt heads, but Bill had a much more cynical outlook than I did. He wasn't shy about keeping the reins on my hyperbole and romance—which was good. He taught me that my reverence for surfing could be a liability. And he was also very good at looking at the big picture."

Surfing's editorial director, Dave Gilovich, brought Sharp and George's ideas into clear focus. Art Director Mike Salisbury framed the work of trailblazing photographers Aaron Chang and Jeff Hornbaker in neon and checkerboard, while budding writers Nick Carroll and future *X-Files* creator Chris Carter further enhanced the brain trust. This was arguably the most forward-thinking, entrepreneurial, and controversial team of media minds surfing ever produced. For better or worse, their focus on young surfers, competition, hot brands, and fashion planted the seeds for what has grown into today's multibillion-dollar surf industry.

However, this transformation required more than style. There were practical necessities, and Collins couldn't help but wonder: If Flame was willing to pay five hundred dollars a month for his forecasts, what about the several million other surfers along the East, West, and Hawaiian Coasts? In 1985, Collins and a buddy named Larry Arnold bought the rights to the toll phone number 976-SURF, and they called their nascent company Surfline. For fifty-five, and later ninety-five cents, any hodad could learn what the waves were doing right then and what they would be doing up to a week out. The fallout from this nuclear bomb is still radioactive twenty-five years on because, in effect, Sean fundamentally changed what it even means to *be* a surfer.

For most Americans, surfers were first revealed in the film and best-selling book *Gidget*. Both were based on the real-life Kathy Kohner and her 1956 summer spent on the beach at Malibu. In one scene from the film, a crush of Gidget's named Kahuna reveals that he's planning to leave the Malibu shack and head for Peru.

"Gotta follow the sun," he says.

"You can't mean . . . ?" she pleads.

"Yeah, I'm a surf bum. You know, ride the waves, eat, sleep, not a care in the world."

She stammers, "Um . . . uh . . . It may be awfully nosy of me, Kahuna, but when do you work?"

"Oh that," says the former air force pilot. "Tried it once, but there were too many hours and rules and regulations."

Before Surfline, Kahuna's outlook was not only perfectly reasonable—that is, to unrepentant surfers—but almost necessary. If you couldn't escape the pull of the waves, you had to make a life of chasing them. But Collins changed the equation. Not only did he take away the surfer's go-to excuse for being jobless, his forecasts made it possible to keep surfing through one's "responsible years." Hell, after Surfline, you didn't even have to live at the *beach*. For big wave surfers, reliable forecasting in the ensuing years would give the unheard-of ability to make a living by launching costly expeditions to meet the biggest storms on the planet.

Surfline was an anathema to Flame and to a great many dedicated local and "feral" surfers who would camp out on a remote spot waiting for the waves to show. Whenever Collins issued a "Surf Alert," cherished empty lineups filled like Interstate 5 on a Friday afternoon. Surfers weren't supposed to give, much less *sell*, forecasts to the masses. It was okay for Flame to plan *his* life around pending swells, but not *everyone else on the planet*. Flame possessed his own considerable forecasting knowledge, but nothing like Collins's. To make matters worse, photographers at *Surfer* could now act on Collins's intelligence. Shouting matches between Flame and *Surfer* lensmen soon ensued at outback spots all along the Baja peninsula.

Collins and Flame's relationship became typically symbiotic yet competitive. If Flame had a good bead on a swell, he might or might not reveal his thinking to Collins. If Collins had the goods (which he always did), he might line up a crew and head to Baja but not tell Flame until he returned with a batch of photos.

Flame kept his building obsession with the Cortes Bank a secret from Collins. Yet Collins had conducted his own recon. His first inkling came around 1988, when a fisherman called to ask Collins about spots along the Channel Islands. In a surfer's quid pro quo, Collins revealed a few secrets, and the fisherman opened up on what he had seen out at Cortes. "He fished and traveled out there a lot," says Collins. "He told me about seeing some good waves at Cortes that seemed to have good shape for surfing."

Collins then ran into Flippy Hoffman down in Baja. The aging charger described essentially the same scene to Collins that he had to Flame.

Yet when opportunity knocked in January 1990, it was Flame who first opened the door. A pinwheeling Aleutian low sent a now-legendary swell charging toward Oahu, creating all-time conditions for the fifty-thousand-dollar Eddie Aikau Big Wave Invitational at Waimea Bay. The swell then steamed toward the mainland with 27 feet of pure, long-period energy. The weather along the California coast was picture perfect. Flame got in a plane and flew over the Cortes Bank, and he told Collins nothing until several weeks later.

Sometime in 2003, Bill Sharp asked Flame about the flight. "A lot of people have asked me that," Flame told Sharp. "I can't really put any one kicker item up on the board leading to the first flight. It was primarily just Mike Castillo being the hellman and the Go-For-It Guy that he is, and me going, 'Hey, well, there's this big swell and do you just wanna go out and look?' He just said, 'Let's go!' You know? It was like having no hurdle to jump over except writing my check for the time we were in the air. It was just someone there going, 'Why not? Let's just go take a look for the heck of it.' So we did."

Early on the morning of January 23, Mike Castillo motored his tiny Cessna along the tarmac at the Oceanside airport and pointed his single propeller toward the naval weapons outpost on San Clemente Island.

"It was funny," Castillo told me in a later interview. "We had to get the okay with the controlling agency at San Clemente—Beaver control—my call sign was November262 Zero XRay. We just played it like we were Coast Guard on patrol out to Cortes Bank, and they were totally cool. I think maybe they knew we weren't, but us sounding official was their way of not barring us from flying through their airspace. We flew right on by off the south end of the island, and when we got out to the open ocean we saw this big navy ship on the water. The lines of swell were fucking unbelievable. The interval on the swell was just *huge*. It wasn't even breaking on the beach because the swell was so deep it was just getting blocked. The real swell never really even got to our beaches. That navy ship, it just looked like a toy boat on those waves."

Castillo brought the plane down to within a couple of hundred feet off the water, marveling at lines of open-ocean swell he reckoned as 18 to 20 feet high. The Cessna buzzed over a school of many thousands of leaping dolphins. Then off on the horizon, their eyes fixed on a telltale hint of white water.

Ten minutes later huge, slow-motion breaking waves appeared through the propeller. "We were screaming at the top of our lungs," said Flame. "It just, it just literally caught us so off guard."

With noses fogging the windshield, and the motor drive on Flame's Canon working overtime, the pair made several slack-jawed passes above Bishop Rock. The sole point of reference was the Coast Guard buoy, which was regularly buried beneath thundering white water Castillo figured was at least 40 feet high. Before the waves broke, they rose high into mammoth slabs perhaps twice that tall. The pair felt like astronauts on Neptune.

"If you surfed down there, there was a serious chance of death or dismemberment," says Castillo. "It was like nothing anywhere else. Even Jaws over on Maui. I mean shit, this is the longest fetch in the world. You could surf a wave out there that had come off Siberia. The potential was unlimited."

"It was huuuge," said Flame. "Finding that—finding it so perfect when no one really had ever gone there. It was truly something that you could have qualified as, 'Wow, we *discovered* something.' You know, something that no one had ever done before, no one had seen and no one else had photographed. It's a really incredible feeling to know that you're the first one to tread out there."

Mike Parsons isn't sure how many days it was after the flight, but he remembers the phone call well. Flame said he wanted Parsons to come down to the *Surfing* offices *immediately*—and he wanted him to come alone.

"Flame had this weird ceremony," says Bill Sharp. "He would usher in the unaware, lock the doors to the photo room, and scare the fuck out of them."

Flame thumbed the lamp to his slide projector. Parson recalls: "He said, 'I have some photos I want you to see.' But before he showed the first picture, he threatened me—made me swear to the utmost secrecy ever. He said, 'Absolutely, if this ever leaves your mouth. . . .'"

Flame keyed the forward button, and Cortes Bank flashed into view. Parsons's chest tightened.

"I just got this wave," Flame said. "It's out in the middle of the ocean. It's way off the California coast, and it's bigger than anything. I've got the boat. I've got the way out to it. All we need is a swell."

Parsons had no idea—no inkling such a wave existed. Flame wouldn't tell him exactly where it was—only that it was *out there*. He showed Parsons all angles—including the shots that showed the buoy in the foreground. Was it 20 or 60 feet? Parsons thought it was probably rideable, but he wasn't sure. What were the currents like? What lay on the bottom? How cold was it? What about big sharks? No one used Jet Skis yet—whether to tow into a wave or for rescue. If you were hurt, well, what could you even *do*?

Parsons emerged from the photo room a little pallid. Sharp gave him a wry grin. "I had mixed feelings," Parsons says. "If I saw that wave today, I'd have been freaking out and ready to go because we have Jet Skis and all the safety gear. But back then all we had was our paddle surfboards. The thought of going out there—it was intimidating as hell."

Of course, such a journey lay well over the horizon. After all, several years still lay ahead before Mike Parsons would be properly—and horribly—introduced to the waves of Maverick's. Yet Parsons, Flame, and Sharp—and soon also Sean Collins—had found their siren song. Out in the middle of the ocean their leviathan awaited—a wave beyond their greatest ambitions and deepest fears.

Chapter 7:

AT
ARM'S
LENGTH

The modern big wave surfer must realize that he wasn't
born with an 18-foot umbilical tethering him to a lithe,
composite gun, and with a detailed lineup chart and
printout of swell predictions in hand. . . . Today's
hyper equipped surfer is the end result of thousands
of years of evolution.

—Dave Parmenter, *Surfer* magazine, August 1999

Eventually, the waves at the Cortes Bank would become an earth-shattering revelation, at least for one particular and rarified surfing subset. Yet the vast majority of surfers blanch at the very idea of seeking out such monstrosities, and truly big waves don't impact their day-to-day surfing lives one iota. That doesn't mean they aren't fascinated, but only from the beach. Most never angle for a spot in those lineups.

One imagines that Ahab must have been similarly lonely, for only a very few of the most fiendishly obsessed captains would have sympathized with and shared his lust for the biggest, most dangerous whale of all. To fully appreciate the impact and lure of Cortes Bank, it helps to understand something of the world of big wave surfing and the treacherous reef breaks that form the sport's crucibles—those places where surfers develop the tools and techniques, and the nerve, they need to approach a place like the Cortes Bank when it is at its height, breaching in almighty, storm-driven rage. Even big wave surfers sometimes forget where they came from, that they "stand on the shoulders of giants," to quote *Surfer* magazine's brilliant and iconoclastic scribe Dave Parmenter. Towsurfing became the big wave surfer's rocket-launched, explosive-tipped harpoon. This bastard mechanized spawn of surfing, waterskiing, and motocross was an evolutionary response to the desire for bigger and bigger waves. When towsurfing arrived, it shook surfing to its core, yet it followed and was made possible only by generations of paddle surfers who paved the way. Before anyone could stand on the shoulder of a giant atop the mile-high pinnacle of the Cortes Bank, scores would suffer, and in some cases die, running the gauntlet on waves whose histories are woven between Hawaii and the North American mainland like a braided necklace of kelp and hibiscus.

Makaha

On a quiet summer morning, Makaha Beach Park seems about as idyllic a place as you might find in all of Hawaii. It's a perfect crescent moon of blond sand and warm, sapphire ocean, hemmed in by a dragon spine of ridgeline that defines the northern boundary of Oahu's arid Waianae Valley. A small scrum of local kids bob atop surf and boogie boards, circling like vultures over the dying remains of a Southern Hemisphere swell. Beneath a small grove of broad-leafed kamani trees, a group of old men nurse coffee and talk story at a picnic table while a crew of teens sit atop a low wall, dissecting the rides of their friends and rivals, and talking shit when a wipeout lets a nice, open-faced wave go to waste. The smiles are broad, the laughs hearty, and the dialect a staccato Hawaiian pidgin, laced with copious profanity.

On a day like this, it seems hard to believe that Makaha could drown anyone, but that assumption would be a mistake. The volcanic seafloor is subject to churning rip currents and waves that can go from playful to deadly serious at the literal flip of a switch. This typically first happens sometime in the mid- to late fall, when titanic righthanders awaken and thunder down Makaha's outside reef.

The history of what we today recognize as the sport of surfing begins in ancient Hawaii, and it reaches back at least seventeen hundred years—well before the eighteenth- and nineteenth-century arrivals of explorers like Captain James Cook, whalers like Herman Melville, journalists like Mark Twain and Jack London, or naval officers like Archibald MacRae. To Hawaiians, surfing formed the basis for an entire culture. In around 1859, a brilliant Hawaiian writer and cultural anthropologist named Kepelino Keauokalani used the recollections of his elders to capture the cultural zeitgeist of precontact surfing in terms that any dedicated wave rider could recognize today:

"Expert surfers going upland to farm, if part way up perhaps they look back and see the rollers combing the beach, will leave their work . . . then hurrying away home, they will pick up the board and go. All thought of work is at an end, only that of sport is left. The wife may go hungry, the children, the whole family, but the head of the house does not care. He is all for sport, that is his food."

In the modern era, the offshoot that's come to be known as big wave surfing is inextricably linked with Makaha, whose formidable righthand waves were plied by countless generations of Hawaiians. The break's first documented charger was the six-foot-six Hawaiian chief Abner Paki, a man who actually made Archibald MacRae's acquaintance during a Honolulu church service in 1845. Paki was reputed to hold off on launching his hundred-pound, fourteen-foot-long koa wood surfboard until the waves were absolutely massive (Paki's board is today part of the permanent collection of the Bishop Museum in Honolulu). Yet the torch of Paki's obsession almost blinked out, along with the rest of Hawaiian culture, when Western disease and oppressive missionaries worked to practically annihilate the islands' indigenous population and society throughout the nineteenth century. The Hawaiian penchant for combining nudity and *he'a nalu*, or wave sliding, was considered the worst sort of godless hedonism and particularly singled out for censure.

Surfing survived only in isolated pockets until the early 1900s, perhaps dwindling to no more than a dozen practitioners at its lowest ebb. After the turn of the century, though, Olympian swimmer Duke Kahanamoku helped lead a revival of the "Sport of Kings," demonstrating and teaching surfing to Westerners and exporting it to the U.S. mainland, where it flourished on California beaches through the first decades of the twentieth century.

In the years surrounding World War II, a handful of U.S. surfers—like Tom Blake, Pete Peterson, Whitey Harrison, Wally Froiseth, John Kelly, and brothers Walter and Philip "Flippy" Hoffman—made the crossing to Hawaii on the promise of surf and adventure. Once here, these Americans went native, leading a feral, carbuncle-covered life so at odds with the suburban postwar idyll that it's almost inconceivable. They were a step ahead of Kerouac and established the rootless, surf-chasing lifestyle that guys like Harrison Ealey and Ilima Kalama would come to treasure. They lived in tents, had no money, and subsisted on fresh fruit and fresher fish. They were also the first Californians to risk themselves on Makaha's frightfully perfect wintertime walls.

The first big Makaha waves of modern times were ridden atop long, dartlike "hot curl" fiberglass-covered balsa wood surfboards that had no fins. They were fiendishly tough to maneuver, but maneuverability wasn't the point. The simple goal was to angle into a wave and head straight for the exit. Walter, Flippy, and an islander named George Downing eventually conducted some of surfing's first serious heavy-water experiments with "skegs," stabilizing fins that would allow them to hold terrifically high and tight lines along the faces of Makaha's mammoths. As the Californians returned to the states and spun their dizzying tales of Makaha back home, they established the break's early reputation as the premier arena where the best surfers could prove themselves in big waves.

Since the 1960s, many of the most gripping exploits at Makaha have taken place under the eagle eyes of Richard "Buffalo" Keaulana. "Buff" is a broad-shouldered full-blooded Hawaiian who can trace his lineage directly back to

King Kamehameha. Buffalo grew up in Waikiki. He was brutally abused by an alcoholic stepfather, and thus spent much of his waking and sleeping life on the beach. But he grew into a gentle giant with a compassionate heart and became a popular diver and champion bodysurfer. Eventually he was named the first lifeguard ever on Makaha—a job equal parts cop, judge, jury, and hellman (a surfer-centric term that at its essence describes a fearless waterman). Buffalo also became one of the best surfers to ever charge Makaha, and he and his fiery, hilarious wife, Momi, raised six children in a tiny bungalow mere feet from his lifeguard stand. In 1976, Buffalo joined the famous cross-Pacific journey aboard *Hōkūleʻa*, a reproduction of a traditional Polynesian sailing vessel. Guided only by the stars, *Hōkūleʻa*'s 2,400-mile voyage from Hawaii to Tahiti proved definitively how Hawaii's Polynesian ancestors sailed to the islands.

Buffalo's oldest son, Brian, is not quite so big as his dad. While he's clearly Buff's son, a sharper facial structure suggests at least a few genes of the haole, Charleston, South Carolina, whalers in his mother's distant ancestry. Brian has left a deep impression on the surfing world. He has won multiple world surfing tandem championships and is widely considered among the most talented big wave surfers Oahu ever spawned. He's also coordinated stunts for films like *Waterworld* and *Pearl Harbor* and is widely regarded as the best heavy-water rescue expert in the world.

On this early August morning, I find Brian and Buffalo posted up beneath a kamani tree that Brian's mother planted in honor of his birth. Brian's charging his batteries with a quadruple espresso, a drink the local Starbucks employees call a "Keaulana Special"; he has already fielded about ten phone calls. Best friend and fellow surfer stuntman Brock Little wants to discuss a job. Filmmaker Brian Grazer might be down for a visit. Brian's seventeen-year-old daughter, Chanel, is going surfing. "Where?" he asks. "Who with? How long going be gone?"

I ask Brian if he worries about raising his two teenage kids along Makaha's shoreline—a stretch still generally referred to as Hawaii's Wild West, and not only for its waves. To Brian, this is the safest place on Earth. "My family is so deep-rooted and connected," he says. "When I was a kid, I couldn't get away with jack. Anything I did, Dad finds out. For my daughter it's even worse. She never realized it till I said, 'Okay, you don't believe me? Go out for the day.' When she got back, I gave her a blow by blow of her day: where she was at class, what path she took to school, where she was holding her boyfriend's hand."

Brian's early years in the water, during the 1960s and early 1970s, were spent under the watchful eyes of the best big wave surfers on Earth—guys like Rick Grigg, Buzzy Trent, Butch Van Artsdalen, Pat Curren, Fred Hemmings, Paul Strauch, James "Chubby" Mitchell, and a wild, burly hellman named Greg Noll. So close were these men to the clan Keaulana that for years Brian thought most were his biological uncles. Noll in particular spent many nights in the tiny family bungalow. Occasionally he awoke from a brutal hangover to find Brian and his brother Rusty grinning down at him impishly from the top of their bunk bed, pissing right onto his face.

Brian remembers watching in awe as "Uncle Greg" challenged the lumbering lines of his home break. "Makaha starts out kind of like humongous Laniakea," Brian says, comparing the wave to another somewhat more forgiving break on the North Shore. "But then you end up at Waimea Bay. But Waimea goes shallow and back to deep—that makes it safer. At Makaha, it goes deep to shallow. You take off on a 40-footer and then it will grow to an 80-footer. It's like running down a hallway full speed and then having the door slammed in your face. I've had the worst wipeout of my life here."

Indeed, Makaha's inside section contains a notorious backwash that can launch you straight up into the air. When Brian was in fifth grade, such a backwash nailed Buffalo and the trip to the bottom broke his neck, leaving him paralyzed for a year. His family considers his complete recovery a true miracle.

For Brian and Buffalo, and the history of big wave surfing, one Makaha day stands out above all others: December 4, 1969. A once-in-a-generation El Niño condition had turned the entire North Pacific into a gigantic tempest. The resulting waves had been hammering Oahu for weeks.

Brian was seven years old, but he still remembers the booming, mist-shrouded waves as they crashed down a half mile offshore and eventually flooded up the beach, threatening to carry his family home into the ocean. Early in the morning, his father paddled out. "It just got bigga' and bigga'," says Buffalo. "Come two, three o'clock, most guys came in. Not Greg."

By late afternoon, the waves leapt up another notch; they were the biggest Noll had ever seen. He bobbed alone, far up around the top of the point with his heart in his throat, well out of sight of his fretful wife, Laura, and of Buffalo. The waves shook the earth and caused the droplets of water on Noll's surfboard to sizzle, as if atop a kettle drum. With daylight fading, and stuck outside of the breakers, Noll accepted the fact that he had one chance at survival—catching a wave to the beach. He reckoned his odds of survival at around fifty-fifty, and yet he had waited his whole life for this moment.

A tremendous set began to feather far on the outside reef. Noll turned to face them and stroked over the first wave, gasping in awe at its power, his adrenalized pulse pounding in his ears, and set his sights on the second. He levered his board, aiming now for shore, and laid every watt of substantial horsepower into his shoulders.

The wave rose up beneath Noll and he leapt to his feet, assuming his trademark wide stance, with his foot near the tail of the board for control. Back on the mainland, Noll described the experience to me. "I'd spent my whole life looking for that one wave," he said. "It was ten feet bigger than anything I'd ever surfed. Ten feet bigger is a leap of faith. You don't fuckin' know what's gonna happen."

Traveling at perhaps forty-five miles an hour, the rushing water amplified even the tiniest imperfections in his board. His skeg began to resonate—humming like a pipe organ.

Buffalo noticed a tiny, airplane-like contrail powdering the misty wall. It was Noll. "It was hard to see because it was near the evening," Buffalo says. "It was a big, big wave."

The wave is long reported to have been between 30 and 50 feet, and undocumented, though Australian filmmaker Alby Falzon claims a hotly debated three-photo sequence of a 25-footer he shot represents Noll's epic ride. It's a debate Noll refuses to enter. Noll simply calls it the biggest he ever rode while Buffalo calls it the biggest he's ever seen ridden—so who's to argue? "Then when he reach the bottom it close out." Buffalo says, "Boom! It just bounce all over him."

"I got the shit kicked out of me," Noll said.

I ask Buffalo if he thought Noll would die. "I wasn't thinking he was gonna drown," he says. "He had good lungs—could hold his breath a long time. Everybody train for these things. Lots of diving in the summer when there was no waves. So when the waves come up, we were ready."

Noll wasn't so sure though. When Makaha gets big, a river of current pours southward, pulling the hapless surfer toward a deadly cheese grater of volcanic reef. Noll struggled while eyeballing the angle he'd need to, hopefully, safely intercept the sand. Buffalo keyed his Jeep and followed along the beach, quietly wondering if his friend would be butchered.

"Buff's just driving down the beach with a six-pack," said Noll. "He's drinking a beer and I'm sailing down the beach. I ended up goddamn near eating it. Twenty feet more and I'd have been fucked. Just before I hit the rocks, I drag my sorry ass out of the water and he hands me a beer."

"Good t'ing you make 'em bruddah," Buffalo told Noll at the time. "'Cause no way I was coming in after you."

It was arguably the biggest damned wave a human being had ever surfed, and that was how it was regarded by the surfing world for the next twenty-five years.

"I came in, I got to the beach, and I was like, okay, for twenty years here, I've been waiting to catch a wave—this sounds very egotistical—but to basically catch a wave bigger than anything ever ridden," Noll said. "That happened. I had waited twenty years, and then what? By the time another twenty years rolls around, I might be in a wheelchair. Usually I'm stoked at the end of a good day of surfing, but it took me two, three days to get out of *that* zone. I just went home with Laura. Here's this guy whose gone through twenty years in the islands, this eager, overly adrenalized monster, and all of a sudden it's over. The pressure was off. The monkey was off my back. It was, well it was like taking a giant shit. I could go surf and just do what I had to do. It was totally unintentional and totally lucky. I could just exit gracefully. A little bit like a prize fighter."

Popular myth is that Noll quit surfing after that wave. He didn't. He just stopped competing and started enjoying surfing more—appreciating smaller days at Sunset and other breaks. After trekking across Alaska, Noll moved to Crescent City, California, and became a fanatical commercial fisherman. He and Laura still live there today.

Waimea Bay

Makaha wasn't the only giant wave Oahu offered. Everyone knew that on those rare occasions when winter swells got big enough, the waves on the North Shore in Waimea Bay were sometimes even taller. Yet like a monster in the closet, Waimea Bay was long considered *kapu*, a place to avoid—too big and deadly to risk.

That reputation was due in no small part to what happened on December 22, 1943. Oahu surfers Dickie Cross and Woody Brown had paddled out to surf the lonely point break waves of Sunset Beach just as the first pulse of a powerful long period swell began to sweep across Hawaii. The swell rose with such ferocity that the terrified young men were unable to paddle against a river of ripcurrent that rises between the breaking waves and the beach on giant swells. Now they faced a nightmare. They couldn't catch a wave and surf into Sunset Beach, and the only possible way they could see to survive was to make the two-and-a-half-mile paddle from Sunset down to Waimea Bay. During much of the year, when the waves are small, Waimea Bay is a calm, paradisiacal cove. When the surf gets big, the bay's deep inside waters make it one of the last places you *might* safely negotiate your way to shore on a surfboard. The men hoped that perhaps they could sneak in between a huge set in the dwindling daylight and negotiate the crushing shorebreak to kiss the sand.

Yet conditions at Waimea were little better. Waves they reckoned at 60 feet were exploding in eighty feet of water across the bay's entire outside reef. They waited and watched, unsure what to do, when eventually a huge set relieved Cross of his board. That left apparently one choice: to try to belly ride a wave tandem on Brown's board. However, another huge wave came, and Cross apparently swam for it, leaving Brown to face another horrific set that stair-stepped out to the horizon. The late, great Brown described the set's first wave to surf historian Malcolm-Gault Williams:

"I'm watching it come, bigger and higher and higher and it broke way outside, maybe four-five hundred yards outside of me. I said, 'Well, maybe I got a chance.' So, I dove as deep as I could go, again, and I just took the beating; a terrible beating. . . . And when I couldn't stand anymore—black spots are coming in front of my eyes—I just started heading for wherever it looked lightish color. You know, you didn't know what was up or down. Wherever it looked kind of a light color, it might look like down, but 'That's where I'm headed for.' And I got my head up!

"So, I figured, 'Man, if I lived through this one, I got a chance!'

"So, they washed me up on the beach. I was so weak, I couldn't stand up. I crawled out on my hands and knees and these army guys came running down. The first thing I said to them was, 'Where's the other guy?' They said, 'Oh, we never saw him after he got wrapped up in that first big wave.' That was their words. 'Wrapped up in that first big wave.' I figured from that, this guy [Dickie] had so much guts, he tried to bodysurf the wave. Because, otherwise he would have dove down. Why didn't he dive down under it? If he got 'wrapped up' meant that he was up in the curl, right? How else would you express it? So, I figured he tried to bodysurf in."

Cross's death put a voodoo hex on Waimea that would last better than a decade. As far as anyone knows, the first effort to ride what might be called the second great big wave on Oahu was led in 1957 by Greg Noll and his buddy Mike Stange. The burly, magnetic Noll and the picture-perfect righthander were ideally suited for one another. On the five to twenty days that Waimea broke each year, Noll and Stange laid down the gauntlet alongside well-known guys like Mickey Dora, Buzzy Trent, Butch Van Artsdalen, Mickey Muñoz, and underground surfers like Harrison Ealey.

Waimea soon became a media sensation, its waves providing fodder for Hollywood films like *Ride the Wild Surf* and *Gidget*. Waimea Bay regulars found occasional, and high-paying work trying to kill themselves standing in for Frankie Avalon and his buddies. Greg Noll rode the movies, photo shoots, and his hand-shaped "Da Bull" surfboards into a rare and wondrous place—an actual livelihood based almost solely around surfing big waves. It's a feat that today, even in a world of million-dollar contracts for postpubescent small-wave aerial artists, is still a rarity.

In the ensuing decades after 1969, a cadre of other surfers would take their cracks at Greg Noll's summit—mostly at Waimea. But despite countless epic rides, the consensus was that no one could, or perhaps would ever, top that wave.

I asked Noll why it was so hard to catch something bigger. "Waves that big are breaking on an outside reef," he said. "They're coming across from the Aleutians, and they're moving much, much faster than smaller waves that are breaking on the inside. The water running up the face makes them even that much faster still. That's why I rode a board that was 11 foot 4 inches long, 22 and ⅞ inches wide, and 4 and ¼ inches deep. That thing was a wave-catching *machine*. If you can't paddle to catch the wave, nothing else matters. But the problem with a big board is that once you catch a wave, you've got this long, banging, slapping piece of equipment under your feet. It's like a bucking bronco. A smaller board is easier to control, but you can't catch waves with it."

In short, there comes a point when a wave is carrying so much energy that you can't match its speed by paddling. The wave passes beneath you or launches you into a deadly oblivion. Waves bigger than Noll's came to exist in a hazy zone that a brash young North Shore local named Mark Foo would coin the "Unridden Realm." After a near-death experience at Waimea Bay in 1985, Foo postulated that to charge into a wave better than 50 feet from top to bottom you would have to be towed onto it from the back of a boat—or maybe a Jet Ski.

The actual origins of Jet Ski–assisted surfing are somewhat murky. But if you were going to place a bet on the very first person to be towed behind a diminutive personal watercraft on a surfboard, the safe money would fall on a man named Wes Laine. Back in the late 1970s, Laine was an aspiring young East Coast pro who spent a lot of time in Carlsbad, California, with his older brother, Randy. Kawasaki tested early incarnations of stand-up Jet Skis in the Carlsbad Lagoon, and Randy, who competed atop both surfboards and waterskis, was particularly fascinated. "I saw the prototypes and used to borrow them and take them out into the surf at Tamarack," Randy said. "One day I tied a rope to the back of one and towed Wes to Oceanside Harbor. Pulled him into, like, 6-footers with a little thirty-horsepower 400 cc ski."

Randy was soon traveling on the worldwide surf circuit on Kawasaki's dime, testing stand-up watercraft in big waves at Todos Santos and running ski demos at surf contests from Huntington Beach, California, to Bells Beach, Australia. He became one of the best stunt riders in the business. "I knew that surfing and Jet Skis were going to come together from the first time I saw a ski," Randy told me. "I just didn't know how it would happen, and I didn't capitalize on it. I just loved doing it so much."

Around the time Foo first pontificated on the Unridden Realm, Randy Laine and a nephew-in-law of Flippy Hoffman's named Herbie Fletcher started riding waves along the North Shore atop a stand-up Jet Ski. They were eventually joined by a good buddy of Brian Keaulana's named Squiddy Sanchez and started charging Waimea Bay and outside reef breaks like Himalayas and Outer Log Cabins on their little one-man aquasleds. At Pipeline, Laine and Fletcher actually towed pros Wes Laine and Brian McNulty, Martin Potter, Tom Carroll, and Gary Elkerton into a few waves. The rides drew cheers but didn't set off the lightbulbs. Perhaps the Jet Ski's loud and smoky two-stroke motor was an anathema to the surfing aesthetic. Perhaps the stand-up craft was too unstable and underpowered. Or perhaps the surf world just wasn't ready to admit that you'd actually need a machine to crack Foo's "Unridden Realm." Randy Laine didn't know it, but one day he and his ski would become intimately, and terrifyingly, familiar with this realm atop the Cortes Bank.

Then, near the end of the second week of 1990, storm warnings sounded from Hawaii to California: a meteorological "bomb" was arriving. A 940-millibar low had generated a fifty- to sixty-knot windfield along better than a thousand miles of Pacific Ocean, aiming 27 feet of deepwater groundswell directly toward Hawaii and the West Coast. Comparisons flew between this swell and the monster of 1969. This would be the swell that Flame witnessed from the air over Cortes Bank.

The waves bore down on Oahu on January 21, and for the first time in four years, the Eddie Aikau Big Wave Invitational at Waimea Bay was a go. Invitees included Mark Foo, Ken Bradshaw, Brian Keaulana, and his good friend Brock Little. The twenty-one-year-old Little was in the prime of his surfing career and felt well nigh invincible.

When the biggest wave of the day came through, Little aimed his gun. The crowd gasped—then shrieked. If there had been a wave comparable to Noll's, this was it. Little could not have positioned himself any better. But that didn't matter. "There was too much water moving," he said. "That wave was not meant to be ridden."

Brock made the drop, but then fell, his life flashing before his eyes as he bounced across the water like a crashing speedboat. The wave swallowed him, but an instant later, he was churned to the top. His head poked out through the roof of the wave, and in the instant before he was buried again, Brock was granted a beautiful view of Waimea Bay, with the crowd and the mountains spread out before him. He wouldn't unseat Noll, but the epic instant before the wipeout nabbed the cover of *Surfer* magazine.

Another incident went down that day, one far less celebrated but no less important. Brian Keaulana caught what he thought was a makeable monster, but he too fell. Brian has a phenomenal lung capacity and the ability to slow his heart rate at will—enabling relaxation under cataclysmic conditions. Still, this was as heavy as it gets. Until he cleared the maelstrom of choking white water in the impact zone, he would be fighting for his life all alone. Of course, that's just how it was. Aside from the occasional helicopter basket rescue, no one had yet devised a means of rescuing a surfer in the absolute bull's-eye of a breaking wave. Then, with a four-story slab of water about ten seconds out, an angel appeared above Keaulana.

"I was getting ready to get pounded," says Keaulana. "Then my friend Squiddy comes right up to me on one stand-up Jet Ski—right in the impact zone. He just zooms in and looks me straight in the eye and says, 'You all right, Brian?' I was just amazed. I was like, 'Yeah.'"

Sanchez zoomed away the instant before the wave detonated, pile-driving Keaulana. But that no longer mattered. "I was just goin', 'That was fuckin' amazing. Someone actually came into the impact zone at Waimea. If I survive this beating, I'm going straight to the dealer, and I'm going to buy one Jet Ski.'

"The next day my wife goes, 'What you do today, honey?' I say, 'I put down five grand on one Jet Ski.' She goes, 'You did *what*?' I go, 'But honey, I'm gonna change lifesaving with this thing.'"

However, when Keaulana tested it out, practicing pickups and rescues with fellow guard and best friend Terry Ahue, he found that lifting someone onto the wobbly machine was very tough. Then Keaulana's mind lit on an old boogie board at home. He poked holes along the board's perimeter and then wove an old garden hose along its length for grab handles before tying it to the back of his ski. "That was the first rescue sled," he says, chuckling. "But it had no stabilizers or anything. I'd rescue people, and they'd be flipping over and over and half drowning. But at least I could get 'em in."

Todos Santos

A truly rideable big wave is a rare and wondrous thing. You need just the right combination of deepwater bathymetry, wind direction, and swell angle to produce a huge wave that you can both get into and escape from. By the mid-1980s, the good big wave spots on Hawaii were, in large measure, already spoken for. You might score big waves if you were from the mainland, but you probably wouldn't score many. This is but one of the reasons that Flame and his compatriots began to search the deepwater nooks and crannies between San Francisco and Ensenada. If they found a spot, they might manage to surf in secrecy—at least for a while—and escape the madding crowds.

After he became a paid forecaster for *Surfing* magazine in 1985, Sean Collins began fine-tuning his predictions. The fiercely entrepreneurial and scientific nonscientist came to recognize a fact that had somehow escaped both surfers and even marine meteorologists. At the true deepwater breaks, it wasn't only the *height* of a swell that was important, but its *period* from crest to crest. The longest-period waves were the deepest, fastest, and carried the most water. If you wanted to find massive surf—and that's all that the growing ranks of dedicated big wave surfers wanted—you found the spot with the deepest offshore water at the very beginning of a swell.

At the same time that they first fingered Cortes Bank as a possible new target, Collins and Flame were lured by the fathomless bathymetry off the Islas Todos Santos, a pair of tiny, uninhabited moonscape islands seven miles offshore from Ensenada. Todos had a few known surfable spots, but everything about it was sketchy. To get there, you drove to Ensenada and then hired an impoverished local fisherman for a chilly, forty-five-minute cruise aboard a rickety panga. The captain puttered around just off the edge of the breaking waves, praying his sewing machine and rubber band two-stroke engine didn't die or seize up amid a blanket of kelp. Due to the lack of accessibility and the fact that no one typically ever reached Todos at the very beginning of a swell, surfers didn't realize how big it could actually get out there. "It all came down to the bottom contours interacting with a long swell period," says Collins. "That's what's key at all the big wave spots—every single one of them."

Collins thought the right off the northernmost island seemed an ideal spot for a long-period wave to rival Hawaii. But very few surfed there, with good reason. It's nasty, scary, brutally cold, foggy, windy, and choked with forearm-thick bull kelp. Just inside the takeoff zone lies a submerged boulder that produces a heart-stopping surge of boiling water just as you're dropping in. If you eat it, you might be pinned against an urchin-and-limpet-lined seafloor or swept across a boulder-strewn shoreline.

In late 1985, Collins saw the right conditions brewing at Todos, and he told Flame to send a *Surfing* contingent that included Bill Sharp, Sam George, Dave Parmenter, and Mike Parsons just as the first long-period waves hit. In the heart of the gladiator pit, Flame filmed Parsons slaying a 25-footer—by far the biggest wave any magazine had ever shown ridden on the West Coast. The caption read, "Sean Collins forecast this swell for Mike and this is what they found."

The wave earned an appropriate moniker: "Killers."

"Boy, that just really set off the sirens," Collins says. "Everyone in California was just like, *Oh my God*. It was the first time anyone realized there was a Hawaiian-size wave just a couple of hours outside of LA."

Subsequent Todos journeys upped the ante. On February 5, 1987, Bill Sharp and Sam George were out when Flame and photographer Rob Brown captured Parmenter, world champion Tom Curren, and pro surfer Chris Burke on even bigger waves. Parmenter later described a moment of sheer, naked panic during the session in *Surfing* magazine: "Missing a wave, I wheeled around to find my companions clawing for the horizon. I went over another, smaller wave, and then suddenly in front of me was a malevolent hillock of water surely sent from the bowels of hell. It was a no-win situation. I didn't know which way to paddle. Easily five times my height, the wave felt bottom, skidded, and vaulted into the lethal slow motion of all deadly things. I felt like I was in the throes of a nightmare."

For latter-day Hawaiian hellmen like Ken Bradshaw, Brock Little, Todd Chesser, and Mark Foo, the photos from this session were a wake-up call. When a swell hit Waimea, the same waves would sweep into Todos Santos two days later. Hawaii surfers began traveling to California for big waves. This was unheard of.

Unlike Hawaii, though, no one lived at Todos. There was no established hierarchy, and no angry locals. All you had to do to prove yourself was to muster up the shriveled cojones to paddle out and make the drop.

Mike Parsons began to hurl himself over the ledge on particularly suicidal waves. Sometimes he would take off on a beastly closeout he knew he couldn't make, just to see how long it would be before he popped to the surface. Thanks to a strange quirk in his genetics, he somehow managed to shake off the steamrolling. He would, in fact, often emerge laughing like a maniac. He scared the hell out of Flame and Rob Brown.

"Guys like Brock or Todd would come over," Parsons recalls. "And I just felt, 'Todos is my spot.' I wanted the biggest wave of the day. I knew every rock and learned everything that happens when you'd get caught inside."

Yet just because you understand the dynamics of being caught inside at one spot does not necessarily make you safer at another. In fact, such overconfidence can make things much, much worse. Parsons would learn this lesson at Maverick's.

Maverick's

During the 1980s, Bill Sharp, Sam George, Mike Parsons, and a handful of buddies also explored, in addition to Todos Santos, a few other bona fide big wave spots off San Clemente and San Nicolas Islands. However, when Sean Collins said it was going to get *really* big, the smart money still lay on Todos.

But there was another spot well to the north. It had exploded in plain sight of Ohlone Indians for eons, just off the vast green headland that marks the northernmost edge of a Northern California hamlet today called Half Moon Bay. Just below that headland lay the perfect tank trap bathymetry to lure in the exact same long period swells that lit up Todos Santos. The wave was first explored back in 1961 by a Northern California surfer named Alex Matienzo, who paddled out with a couple of friends on a 6-to-8-foot winter's day. His gutsy white German shepherd kept following him out to sea, so eventually Matienzo locked him in his car. The dog's name was Maverick.

In 1975, a young third-generation Half Moon Bay local named Jeff Clark paddled out to Maverick's for the first time. He was awed by the water's godlike power and menacing emptiness. He tried to convince his friends to make the mile-long paddle out to surf with him. But no one else would dare take the drop. It was too damn scary.

On a cold boat ride back from the Cortes Bank in November 2010, I sat alongside a pair of rapt young hellmen named Greg Long and Mark Healey as Clark described his first ever paddle out in 1975.

"So I get out there and there were long lulls, and I see this set coming and I'm way too far inside. I just start scratching, *just* getting over these lefts that are bowling and breaking. And then I get way outside and I say, 'Okay, here's where the wave breaks. Here's my landmarks.' I got in position for the next set and paddled with it, trying to feel the energy and just trying to find that vein to get into a wave. Once I did that, it was like, okay, no looking back, you're going. I felt like him today [points at Greg]. Only on a smaller scale. So I paddle into this thing and I just remember it humping up like our beach-breaks. I got to my feet and I just remember the shadow behind me. I'm just running straight, like, frickin' *don't get me*. 'Cause you *know* the lip's coming. And I made it. I rode five waves that day. Never had a wipeout. After that, it's like, you can actually ride this wave, you know? I'd seen it on much bigger days, and I was just like, *It's on*."

For fifteen years, rumors of the wave and its mystical surfer ebbed and flowed. Through all those long winters, Jeff Clark surfed alone, accompanied only by whales, sea lions, curious otters, and big, toothy fishes. It wasn't until January 22, 1990—a day after Brock Little and Brian Keaulana's epiphanies at Waimea Bay and a day before Flame and Mike Castillo's jaw-dropping first flight over the Cortes Bank—that Clark lured a pair of Santa Cruz buddies, Dave "Big Bird" Schmidt and Tom Powers down to Half Moon Bay. "I said, 'You guys wanna see a perfect peak?'" Clark told *Surfer*'s Ben Marcus, "'Come with me.' We snuck off to Mav's and walked to the top of the lookout. Schmidt was looking off going, 'Where is it?' and just then a set came through. Big Bird started pacing back and forth going, 'Oh my God!' Powers was going 'What? What?' And Schmidt said, 'That's Waimea.'"

But it wasn't Waimea. It was something even scarier.

An aspiring journalist named Evan Slater, a pair of brash loudmouths named Peter Mel and Ken "Skindog" Collins, a tall, noisy oncologist from the Bay Area named Mark Renneker, and a quiet young man named Jay Moriarity became part of an expanding crew who took Clark up on his invitation and began sharing the waves. One cold and gusty morning in December 1994, Slater sat alongside Moriarity as he paddled for a solid 30-foot bomb. "As he started paddling for that wave, I just said, 'Good night, Jay,'" Slater recalled.

Surfer lensman Bob Barbour's motor drive clicked through a rapid-fire sequence. The instant Moriarty stood, his 10-foot 8-inch Pearson Arrow was lifted from beneath him, the offshore wind flicking him skyward like a speck of dust. The only part of Jay visible in the ensuing *Surfer* cover shot are his arms, flayed out and flapping hard, while his board is aimed directly skyward. He hovered, impossibly, almost majestically, in midair for an eternal second before being launched into a bone-crushing two-wave hold down. The horrible moment was dubbed "The Iron Cross." It would come to be recognized as perhaps the most horrific wipeout ever caught on film—much less survived.

A couple of days later, with the swell winding down, a trio of Hawaiians, Ken Bradshaw, Brock Little, and a recently engaged Mark Foo, decided to catch the red-eye from Honolulu. Only Bradshaw had surfed the wave before—at a smaller size—and all wanted to see if this wave was really worthy of comparisons to Waimea Bay. Fellow Maverick's virgin Mike Parsons joined photographer Rob Brown on a coffee-charged drive up from the San Jose airport.

Jeff Clark stood in reverent awe early the next morning. Four of the best big wave surfers on Earth would be consecrating his home break with him. Yet none of these men were immortal, and the waters around Maverick's were anything but holy.

Parsons found Evan Slater and a small crew of surfers loading onto a boat called *The Deeper Blue* and was invited aboard. He was wonderstruck by the bluebird conditions—and the wave. He had studied photos and video of Maverick's—but not as hard as he should have. Nor did Little or Foo spend a great deal of time talking to Jeff Clark about hidden perils and currents. But that was okay. Somehow, they always popped up.

Rob Brown climbed up the steep, muddy cliff above Pillar Point, scared out of his wits that he was going to slide down and die. The sun shone through broken clouds, and a light breath of offshore wind left a twinkling mist behind massive green righthanders. "Other than Waimea, I'd never seen anything like it," Brown said. "Taking off on a wave there was like free-falling off a cliff."

Parsons immediately and successfully scored a couple of epic waves. "I was screaming and yelling, and we were having a ball. Mark Foo was having a blast. We were laughing about leaving our girls behind on the day before Christmas, and just saying, 'Wow, isn't this just so great? There's all these great waves in the world, and we get to ride them.' Mark was just on fire. It was such a small crowd—six to eight of us. It was just so bitchin'."

At just after 11 A.M., the horizon turned a deep emerald. The small pack of surfers clawed the water. The lead wave was relatively small, and it was allowed to pass. Foo and Bradshaw were in position for the second. Bradshaw actually had the inside line—the unspoken rule among surfers is that whoever is closest to the breaking curl has the right-of-way. But Bradshaw saw that Foo was a few more strokes into a commitment. He grabbed the reins, wrenching his board into a vertical position to halt its momentum. Foo dropped down a medium-size wave—perhaps 25 feet from top to bottom. It jacked to vertical in an instant.

Approximately fifteen seconds behind came another wave—a near carbon copy of Foo's, but 15 percent bigger. Parsons paddled like hell. In his peripheral vision, he caught a glimpse of Brock Little. Both popped to their feet almost simultaneously, bent low, and negotiated a zero-g drop. But at the bottom of the wave, weightlessness gave way to a high-g compression of legs and body. Parsons lost his balance and leapt feet first off his 10-foot 6-inch Timmy Patterson. A half second later, Little realized he had no choice but to straighten out and leap, too.

Resistance was futile.

"The lip just crushed my chest," Parsons said. "And I was immediately way down on the bottom."

The shockwave was an invisible fist that held Parsons prostrate forty feet down. His eardrums nearly burst and the ice cream headache brought on by the freezing water was horrendous. Suddenly, he was stunned to feel Little bang into him from *below*. They became briefly entangled by either kelp or a surfboard leash. Beneath the foam, Parsons was completely blind. Little didn't seem to be struggling, just sort of bouncing against him like a wayward balloon. Mike imagined a dire scenario: Little might be unconscious. If you survive, how will you rescue him—down here?

Another wave dragged Parsons away by his ankle leash like a cowboy with a boot locked in the stirrup of a runaway horse. After nearly a minute in swirling, frigid blackness, he clawed to the surface in a near panic. Little had been swept much farther toward shore.

"It was so obvious to me when I was down there that I had been wrapped up with Little," he said. "When I saw him, it was just so relieving. All my fear just instantly went away."

Parsons and Little believed that they could paddle north and make it back outside. They dug hard for thirty seconds, but a river of current swept them toward a limpet-and-mussel-encrusted series of boulders. Their leashes were snagged on the bottom as successive walls of white water waylaid them. The torrent was so strong that neither could even bend down far enough to release the velcro on his leash, and both commenced drowning anew. Parsons was stuffed under a rock. He struggled mightily, then tried to remember the instructions Brian Keaulana gave him for just such a dire situation: Relax. Think. But the current didn't let up and neither did he. He escaped from beneath the rock, but his leash held fast. A curtain of blackness closed across his peripheral vision. Blood cells were dying, and his brain was starving of oxygen. Suddenly his leash inexplicably released. Jolted with fresh adrenaline, Parsons groped for the surface. Little had been miraculously freed, too.

Parsons was groggy, but he seemed to recall that the first thing he asked Little was something to the effect of "how radical was that?" Little agreed. Parsons assumed that Little understood the question to also imply that he was amazed when Little bounced off him. Amazingly, Little paddled back out. Parsons pointed his beat-to-shit Patterson out to *The Deeper Blue*. Filmmaker Steve Spaulding pressed his record button.

> Parsons: I thought for sure I was dead. I don't know how I lived.
>
> Spaulding: What happened to Brock?
>
> Parsons: He was drowning, too. . . . I felt him come up underneath me. I felt him, like, banging underneath me. Then we both got thrown, right in the middle there, got thrown over all those rocks.
>
> Spaulding: That was quite the intro to Maverick's.
>
> Parsons: Wasn't it, though? That was by far and away the raddest thing that ever happened to me.
>
> Spaulding: So where's Foo? Foo broke his board.
>
> Parsons: I don't know. I thought me and Brock—we're both in the rocks going, we're gonna die in these fucking rocks. We couldn't come up.

An hour later, clouds darkened sky and mood, and onshore breezes signaled the end of the session. Evan Slater paddled over, and *The Deeper Blue* began to motor back toward the baleful mechanized drone of the foghorn at Pillar Point Harbor. A few minutes later the crew spotted the bottom third of Foo's yellow-and-purple surfboard. Then a black shape lazily lifted into view on the crest of a swell. Dread washed over Slater and Parsons with a force greater than any wave. It was Mark Foo, who'd drowned unnoticed on a wave no one thought was a killer. Everyone had thought that, after his wipeout, Foo must have paddled in. Instead, his body was pulled aboard *The Deeper Blue*, which sped back to the harbor.

When Rob Brown skidded down the cliff to find paramedics performing CPR on a person sprawled across the hood of a car, he thought it was Parsons. He then watched, ghost white in shock, as they zipped the yellow bag on Foo's body. Sobbing uncontrollably, Parsons forever abandoned his surfboard and drove back to the San Jose airport with Brown. "Mike's on the pay phone with Flame bawling his eyes out while all these people were going to celebrate Christmas," said Brown. "It was overwhelming. Surreal."

Jaws

The North Shore of Oahu is a tough place to be a little boy. Grommets grow up playing chicken with a shorebreak that can snap their spines in the blink of an eye, and they dig sandcastles in front of wicked rip currents that can drag them to their doom before Mom even realizes anything is wrong. When a local kid starts to surf, he enters a *Jungle Book* meets *Lord of the Flies* world where the bigger kids egg the little ones into maulers that they have no business surfing. Yet if a kid doesn't at least attempt the drops, the words, taunts, or fists can hurt even worse than an impact with fiberglass sticks or coral stones.

A big little kid named Laird Hamilton came of age in this palm-shaded gladiator arena. He was sucked out in vicious rips and rescued by his adoptive dad, Bill Hamilton, more times than he can count. Alongside his dad, he was probably one of the youngest kids of his era to surf Pipeline and big Sunset. I once asked him if there was a particular instance—a critical moment early in life—when he realized he was out in far bigger surf than he should have been, and if the moment taught him anything.

With nary a pause, he told a story I'd never heard: "I was thirteen years old, and it was at Hanalei [Hanalei Bay on Kauai]. It was when I went from riding what would be considered normal surfing conditions to what we would classify as big. I wiped out three times in a row. That was before we had good leashes. The first wave I caught, I got nailed. Probably swam a mile. Back out, same thing again. Back out, same thing third time. I remember saying to Bill, 'Man, I don't know what's wrong. I just got worked three times. I was hammered.' He said to me, 'Well, now you know, at the worst, what's going to happen. Now go out and catch one.'

"The next one I made—then the next. That was the decisive moment in my psyche. I could survive a pretty good beating and I didn't get discouraged. If I had gone in, maybe I wouldn't have pursued surfing as I have. That was the psychologically defining moment for me. You go out and you have grown men scared—big wave riders scared. You're thirteen years old and you decide you really like this. That it's what you've really wanted—what you've dreamed of."

Laird became known as a preternaturally talented paddle surfer. Had he ended up a little more lithe and half a foot shorter—the size of a Mike Parsons —he might have become a serious contender for the ASP World Title and its world of smaller wave venues. But Laird developed the kind of height, physique, and ego that put him in the company of men like Greg Noll, while his Nordic good looks made him impossible for the fashion industry to ignore. Big waves and bright lights were his destiny.

In the middle of the tempestuous, El Niño–fueled winter of 1991–92, Hamilton and his friend Buzzy Kerbox, a sailboarding champion, began experimenting with a forty-horsepower inflatable Zodiac boat, a waterski tow rope, and a big wave surfboard at a series of mysto reefs off Oahu's North Shore that included Outer Log Cabins. Within sight of some of the most insanely packed and dramatic surf spots on Earth, the duo used the tiny boat to tow each other into gigantic, dreamlike waves all by themselves. If they had been paddle surfing at Waimea, they might have caught five or six waves all day. Instead they round-robined continually, riding more big waves that day than they typically would in a year.

Their session went mostly unseen by surfers, who were hustling for waves close to shore. But a lifeguard named Darrick Doerner watched spellbound through a pair of binoculars. These guys were on to something, and he wanted in.

The following winter, Doerner joined Laird and Kerbox on Maui, at a spine of reef off the winding road to Hana that had the uncanny ability to turn northwest swells into perfect monsters. The spot's Hawaiian name was Peahi, but the surfers decided to call it "Jaws."

Laird had paddle surfed Jaws at small size a few times and was sufficiently intimidated. He said, "Anyone that's ever come to Jaws, the one thing they always comment on is, 'Omigosh, the thing is moving *so fast*. It's just moving at a different gear. Other waves are third gear, others are fourth. This one is OD—overdrive."

Mechanized surfing seemed to hold the most promise. Laird and a small, tight circle soon ventured out in the following configuration: One or two surfers would pilot single stand-up Jet Skis for rescue. Another pair drove the inflatable boat and slingshotted a surfer into the waves from behind. It was frightfully dangerous. The Zodiacs possessed not only sharp propellers but a disquieting tendency to launch backward into the air from the stiff wind blown off the top of a wave. Still, the team managed to ride and study the wave and narrowly survive trips through its wormhole barrel. Jaws became an absolute—and largely secret—obsession.

Sometime around 1987 the first two-person, sit-down personal watercraft—the Yamaha WaveRunner—hit the market. Hamilton and his friends got their hands on one in the early 1990s and called it a turtle. It could carry more than one rider. It was fast, stable, reliable, and most importantly, one of Brian Keaulana's grab-on rescue sled inventions might easily be mounted to its rear. Freed of the buoyancy needed for paddling, surfboards were shorn down dramatically in dimension and fitted with chop-defeating lead weights and windsurfer foot straps. Now, with feet locked in, the largest waves anyone had ever seen were not just ridden by Laird and his friends, they were *ripped*. Massive top-to-bottom carves, impossible aerials, and gaping pits became commonplace across a mind-blowing blue water playing field. It was a quantum leap in performance—a completely new way to inject the earth's purest form of energy directly into your veins.

The images of these revolutionary exploits hit the stunned staff of the surf magazines like a gut punch. In 1994, *Surfer* associate editor Ben Marcus sat in his San Clemente home watching advance copy footage from these epic Jaws sessions for the first time. He keyed his remote back and forth, replaying the titanic rides and rodeo backflips in sheer disbelief, continually repeating, "This changes everything."

Heated debates ensued among Marcus and *Surfer* magazine's other editors Sam George and Steve Hawk. Was this even *surfing*? If it wasn't surfing, then what the hell was it? And what did this mutant—or what some called satanic—hybrid of snowboarding, waterskiing, and monster truck racing mean to the future of the sport?

The contrasts were stark. Into surfing's longstanding, environmentally pure world of rugged individualism was born a terribly polluting, technology-driven team dynamic. Early skis in particular burned and belched a nasty, pungent mixture of gas and oil.

In big wave lineups, the rule had always been "survival of the fittest." You had no choice but to earn your place among the world's heaviest swells—or drown trying. Few begrudged the added safety that Jet Skis brought to the lineup, particularly in light of Mark Foo's death, and increased safety might allow paddle surfers to push even harder. Yet something was also lost with the addition of the ski—particularly with the rise of towsurfing. Whatever this new sport was, it heaved time-honored traditions of self-sufficient, primal forays into the wilderness out the window. Big wave surfers were no longer Lewis and Clark. They were astronauts.

"You gotta match power with power," Laird told me. "Or as Darrick Doerner says, 'Horsepower with horsepower.'"

K2

In the years just before horsepower-driven surfing hit the mainstream, Bill Sharp and Sam George grew weary of editing *Surfing* magazine. Under their tenure, the magazine had grown from just over 100 to 250 Day-Glo pages. "But it got to be like Mexican food," Sharp said. "Same ingredients, different mix."

George made the move over to *Surfer*, but in an unusual move for a journalist, Sharp saw opportunity in a storied brand of surf trunks, "Canvas by Katin." He and his best friend Rick Lohr convinced owner Nancy Katin to put them in charge of wholesaling, and they took the company from a literal cottage business to one with fifty-three employees and millions in sales. In 1997, Katin sold to the mountain sports juggernaut, K2. Sharp stayed on for a while to help direct their marketing while using some of the money from the sale to launch a hilarious gossipy tabloid titled *Surf News*.

Late in his Katin reign, Sharp came up with an explosively controversial idea. Sometime in the late 1950s, Waimea pioneer Buzzy Trent famously said, "Big waves aren't measured in feet, but in increments of fear." Big wave surfers were notorious for applying a weirdly inverse, macho nobility to the actual measurement of waves. Mark Foo would look you straight in the eye and call a giant Waimea Bay wave that clearly measured 40 feet from trough to crest only 20 feet tall. Ken Bradshaw might then scoff and say, "Shoots, it was 18 feet tops." Even more confusing was a Hawaiian tendency to measure a big wave from its *back*. Who the hell rode the *back* of a wave?

Sharp thought this whole sensibility completely ridiculous. It not only made it impossible to measure how big, exactly, a wave was, but this strange logic had the effect of reducing the importance and significance of a big wave surfer's accomplishments. No climber ever underestimated the height of K2 or Mount Everest. Sharp's attitude was: Don't tell me how big you *think* the wave was. Let's cut the bullshit and measure the damned thing from top to bottom— and may the best man, or woman, win. In fact, why not reward the surfer who rides the very biggest wave *a thousand dollars a foot*? A 40-foot wave would pay a surfer $40,000, a 60-footer, $60,000, and so on. . . .

Sharp took his idea to Quiksilver, Billabong, and several other companies, but strangely, no CEO in the surf world wanted to endorse such a radical competition. He turned, instead, to the most aptly named new surf company on Earth, and the K2 Big Wave Challenge was born.

"In a sport of '*men who ride mountains,*' Bill created Mount Everest," Sam George said. "He put a number to the size. He single-handedly changed the scale of how surfers measure big waves. It was also a concept that the mainstream media—not to mention mainstream America—could easily grasp."

K2 happened to endorse Sharp's idea during the epic El Niño winter of 1997–98. A panel of surf industry icons agreed that one of Mike Parsons's best friends—a respected pro from Carlsbad named Taylor Knox—had narrowly defeated Maverick's icon Peter Mel after paddling into a 52-foot beast at Todos Santos. Knox deposited his $52,000 check at an ATM. "I got more publicity for that one wave than Kelly Slater did for any of his world titles," he later said.

After a two-year hiatus, the K2 Big Wave Challenge would change its name and morph into an annual contest, eventually called the XXL, which would come to include yearly awards not just for the biggest ridden wave, but for the worst wipeout, the biggest barrel ride, and simply "Ride of the Year." For decades, the professional World Tour had been the only ongoing annual crucible through which surfers made their reputation. The XXL would become, and today remains, the standard by which big wave surfers are measured.

Interestingly enough, however, Taylor Knox hadn't ridden the biggest wave in the winter of 1998: that honor actually belonged to Ken Bradshaw, who rode a wave at Outside Log Cabins on Oahu that was variously estimated at between 60 and 80 feet on its face and would for some time be known as "The Largest Wave Ever Surfed." The difference? Knox paddled for his 52-foot wave, while Bradshaw was towed into his. The K2 contest was only open to paddle surfers, so Knox got the money. Yet the fuzzy footage of Bradshaw's giant had given everyone—particularly Bill Sharp—a glimpse of the future.

Out at Cortes Bank, the giants slept while paradigms shifted. At least two missions had been scrubbed in the early 1990s due to mechanical issues. Then in 1995, Flame recruited Mike Parsons and Brock Little for a paddle surfing mission on a promising swell. But as their boat rounded San Clemente Island's Castle Rock, a wave-shredding gale sprang to life. This would be the last attempt at the summit for six years.

Flame thought it just as well. "Evan Slater, he's thoroughly convinced that if there had been a pure paddle surfing trip, someone would have caught some waves," he said. "I'll betcha, too, but someone would have paid . . . dearly. I don't know what we were thinking really. We didn't have any rescue craft available. I was gonna shoot from the water—from a surf mat, or maybe a Morey Doyle [a soft foam longboard]. It would have been really difficult. There just was always something in the way, and it's all the better we didn't go. I think it was the Lord making sure we didn't go out there."

One day in early 2000, as Bill Sharp thumbed through a batch of impossible paddle and towsurfing shots, another lightbulb went off. Perhaps this towsurfing could provide the tools and horsepower for a successful return to a place most surfers in the world had never even heard of. He imagined a serious, high-dollar expedition to the Bank. Four teams of the best towsurfers in the world, a five-thousand-dollar contribution per sponsor, a small fleet of skis, a hundred-foot yacht for a base camp, and major coverage in the mainstream media from CNN to the *New York Times*.

Sharp brought the idea to Flame and *Surfing* publisher Bob Mignona. Both men liked what they heard.

Sharp typed up a press release in his prototypically glib, hyperbolic style—something to drum up a sponsor's interest:

PROJECT NEPTUNE!
    COMING TO A SURF SPOT . . . FAR FROM YOU!
    Remember the K2 Big Wave Challenge? It was the year of the El Niño, when the waves were good and the water was warm. The K2BWC issued a challenge . . . One wave, one ride, biggest wave of the winter that was successfully ridden in the North Pacific wins a huge cash prize. That was possibly the sport of surfing's most publicized media event, which resulted in plenty of ink & air-time for the sponsoring K2 Corp . . .

    OK, now for the sequel to the K2 Big Wave Challenge. It's called: Project Neptune. From the brain trust of William 'Bill' Sharp, a.k.a. Dr. Evil, comes yet another surfing challenge the world has never seen, or imagined.

    A journey into the unknown, the abyss . . .

Chapter 8:

THE

PRISONERS

"Who ain't a slave? Tell me that."

—Ishmael, from Herman Melville's *Moby-Dick*, 1851

In some way or another, every serious big wave surfer alive today is a walking ghost. Each one I've met has been absolutely sure at some point that he—or she—was going to drown. Most remember the experience vividly, from the panicked groping for the surface right down to the eerily peaceful point when a hypoxic cloud darkened their vision. Not even a water-safety guru like Brian Keaulana is immune. This is how he described a near drowning after voluntarily paddling into huge surf at Sunset Beach, wiping out, being folded across the rail of his surfboard by the downward force of the wave, and having nearly every molecule of air driven from his lungs: "There's this feeling of black velvet being drawn. Like black drapes in front of your eyes—but your eyes are wide open. Then you start going through twitches and convulsions. Your body's just going through this eruption—and you have no control over it. I was thinking of past, present, future—how I would be buried—and my ashes being spread in the ocean."

Had Keaulana actually blacked out, his larynx would have, in all likelihood, relaxed, and he would have inhaled better than a liter of ocean. The bronchial immersion would have rinsed the thin layer of gas-transporting surfactant from his lungs. So even if he had bobbed to the surface and been rescued, recharging his blood with oxygen would have proven most difficult. As it was, though, the reality of actually dying shocked him with adrenaline, and vision and control briefly returned. He squeaked to the surface and gasped the sweetest breath in his life.

Despite regular moments like these, big wave surfing is, statistically speaking, a relatively safe endeavor compared to other extreme sports. Not counting those who have died of freak accidents like breaking a neck in smaller, near-shore surf, you could count on one hand the number of surfers who have died in truly giant waves during the last two decades. In contrast, and relatively speaking, high-altitude mountain climbers and motocross riders are killed or paralyzed with sobering frequency.

Back in 2002, I interviewed a young hellion named Travis Pastrana for the *New York Times*. By the age of eighteen, Pastrana was well on his way to becoming a sort of Kelly Slater of motocross. Yet while Kelly has endured cuts, contusions, sprains, and the occasional concussion during the course of his world title–filled career, he has not endured anything like what Pastrana had been through by his eighteenth birthday, including thirty broken bones, twelve surgeries, and ten concussions (one concussion alone can cause brain damage). When Pastrana was fourteen, he actually dislocated his spinal column from his pelvis. "Every time I've gotten hurt," he said, "it has been worth it, and it has been my fault. After I separated my spinal column and woke up a week later in intensive care, the first thing I remember my mom saying was, 'Are you sure this is worth it?' There was never a doubt in my mind."

Pastrana's seemingly insane devotion to such an apparently self-destructive activity is something most of the big wave surfers I've ever met have shared. It's the same devotion that drives mountain climbers to want to summit Mount Everest despite grave personal risk and a complete lack of control over the elements: It is all too easy, despite sherpas and oxygen tanks, to get caught in a blizzard and freeze to death or lose fingers and toes to frostbite. Indeed, in 2010 alone, four people died on Everest while 513 reached the summit—thus making one year on Everest more deadly than all of big wave surfing in the last decade.

A number of theories have been raised over the years about why big wave surfing has such a comparatively low fatality rate. One is that unlike sports such as motocross and high-altitude mountain climbing, surfing traditionally has never depended heavily on machines to carry them into waves or on gear as a safety net. Technology, in fact, comes with two flaws: It can fail, and it can fail when it's carried an athlete much further than they'd have gone on their own, leaving them exposed and vulnerable. Consider the Himalayan climber who runs out of oxygen at 28,000 feet just as a storm rolls in. Some have argued that because a surfer isn't insulated by such technology, they're better at staying within their limits and thus big-wave surfing is much safer than it might appear.

This idea was raised most recently in a spring 2011 story in *Surfer's Journal* called "Death Trip" by former *Surfer* editor Brad Melekian. In it, Matt Warshaw, a former professional surfer and the author of *The History of Surfing*, is quoted as saying: "Surfers, when you think about it, have always had a lot invested in the idea that what they were doing was deadly. In reality it's not that deadly at all." Without question, big wave surfing, as a sport, enjoys thinking of itself in such life-and-death terms; this helps burnish its heroic profile. Yet I respect-fully disagree with the implication that, beneath the bluster, big wave surfing is somehow actually safe, or at least inherently safer than other extreme sports. Despite what the statistics say, the constant threat of dying is very real.

In addition, towsurfing with a Jet Ski has blurred the once-fundamental distinction between surfing and other motorized or gear-intensive sports. Obvi-ously, in a sport like motocross, the machine *is* the sport, and the machine itself is what's most likely to lead to your shattered body or death. In big wave surfing, the Jet Ski might hurt you, if it runs you over, but its main function is to throw you into the most dangerous waves possible. The Jet Ski is also a rescue vehicle that can yank you out of harm's way, but if you wipe out and are held down, you still have to reach the water's surface on your own. If you hit a reef and get knocked out or are stuffed beneath a rock, even a life vest won't do you any good. And a life vest only increases a margin of safety, anyway. Typically, only towsurfers wear them at all—since paddle surfers find they make it too difficult to paddle or dive under a wave—and a big wave, a *really* big wave, can submerge even a life-jacketed surfer for minutes at a time. In short, even with every safety measure in place, a surfer can still easily drown. And in my mind, short of being burned alive, drowning in big waves is about the scariest thing that can happen to a person.

In fact, to me, this fear is what sets big wave surfing apart and makes the difference. In terms of fatalities, I believe the statistical discrepancy has less to do with the sport's actual level of danger than with its inherently terrifying nature. Of course, motocross riders and mountain climbers have their moments of terror, but the very ability to surf is predicated on learning how to cope with the simple, primal fear of suffocation—of drowning. The central skill sets of surfing, those basics you need to even catch a wave and stand on a surfboard in moving water, take years to master, and they include countless near-death lessons along the way to scare you straight. Initially, even the least-critical conditions humble the novice surfer, and through these experiences he or she develops the astonishing myriad of physical and neural strengths necessary to negotiate a bigger wave. But as important, this incremental skill-building, and the accompanying brushes with mortality, usually leads most surfers to con-clude that they don't need the heaviest waves to be happy. This fact alone keeps most surfers alive. But then for those surfers who continue into larger and larger waves, they steadily learn how to control their emotions and survive what would otherwise kill them if they panicked.

Let's look at it this way. Say you've been surfing hard for better than a decade. From Santa Cruz to Sunset Beach, you've challenged—and been occasionally soundly beaten by—sizable waves, working you're way up from knee-high peelers

ers to solid 8- to 10-foot, or what most surfers would call double-overhead (a wave twice as tall from crest to trough as the average six-foot-tall surfer). You think, maybe, you're ready to give Maverick's a try—on a very small day. Of course, small at Maverick's is relative. A wave has to be 12, 15 feet—almost triple overhead—to even break. Still, when a mid-period northwest swell hits the California buoy, you swaddle yourself in neoprene and wax up beneath the dramatic headland at Pillar Point. The paddle is long and spooky, but deceptively easy, and soon you find yourself sitting just outside a small, tightly clustered pack of laughing, trash-talking Maverick's regulars. Some of the best big wave surfers alive are taking drops on waves 15, maybe 20 feet high. Among them are Jeff Clark, Mike Parsons, Peter Mel, Ken "Skindog" Collins, Darryl "Flea" Virostko, Greg and Rusty Long, and Evan Slater.

You spend half an hour in deep water off to the side, just watching, sizing up the waves. Then another set appears, and you take a position slightly to the inside and right of the crew. You hope to ride one of the smallest waves. Just take off on the shoulder. Soon enough, such a wave comes and the regulars let it pass. You line up and dig with all your might as it swells beneath you, but you've never felt a swell move so fast or with such power. The wave rolls past you, preventing you from catching it but allowing you to narrowly dive off the back. You kick and stroke backward as hard as you can to avoid being launched over in the wave's pitching lip—a narrow escape. Your board, though, is carried over the falls, giving your ankle leash a long stretch and your knee a solid yank. You feel it pop, but you're okay; you don't get sucked over and into the white water. That was fricking scary.

Just as you finish reeling in your board, whistles come from the pack. A much bigger, more westerly set is stacking up. It loads up farther along the more southerly edge of the reef, and there's no way you'll escape it. The veterans dig confidently for the horizon, and you're granted a filmmaker's view as Peter Mel makes a picture-perfect drop down the first wave. It then falls on you—a two-and-a-half-story wall of liquid bricks that nearly rips your limbs from their sockets and sends you churning beneath black foam. Your neoprene hood is blown off, inviting the frosty, forty-nine-degree water to squeeze your skull like a vice of solid ice. The feeling is utter powerlessness and stark, airless terror—a come-to-Jesus specter that makes you realize you know nothing about the ocean or surfing and, my God, your children—what were you *thinking*? You pray you'll be able to reach the beach and kiss the sand and your wonderful, beautiful wife and kids again. If God will grant that, you'll never do anything so stupid again.

This is precisely why only the most practiced or hell bent repeatedly head out into truly giant surf, and why once out there, they tend to survive. The typical surfer at Maverick's possesses the breath-holding ability of an apnea diver, the flexibility and focus of a yogi, the strength and endurance of an ironman, and the guts of a gladiator. In short, the lineups have already been strained through a .5-micron human filter. Everybody out there knows, once you paddle for any wave and get caught inside and pushed down deep, you can't rely on a Jet Ski to save you. As you progress in surfing little by little, you decide your limit, which for 99 percent of surfers is something far less lethal than a smallish day at Maverick's.

In this hypothetical scenario, you are very lucky to surface with little more than shattered pride and a tweaked knee, and you know it. In big waves, every hold down can simply become a fight for your life. Water is of course softer than rock, dirt, or trees, but when you fall from four stories up on a wave, you might as well be landing on cement. That's why it still remains something of a mystery even to big wave surfers, why more don't die or become critically injured—because when things *do* go wrong in giant surf, they tend to do so in a dramatic and awful fashion: faces peeled off on coral outcroppings, spines snapped, bones blasted through skin, muscles hammered into jelly, eardrums bowed inward until they burst. Medium-size Maverick's has nearly torn off Evan Slater's right leg, and it once pretzeled big Peter Mel so hard that his feet slapped the back of his head. The simple impact of a wave in 1997 left Ken "Skindog" Collins certain that he'd broken his back. In 1994, a simple, straight-up Maverick's wave held down and almost simultaneously drowned Mike Parsons and Brock Little. Moments later, of course, it *did* drown Mark Foo.

So what is it that drives that 1 percent of surfers to willingly and regularly risk life and limb, whether aided by a Jet Ski or not? Why is it, not merely that they do it once or twice, but that they are so often driven to keep surfing big waves that regularly subject them to hold-downs that would scar most people for life?

Of all big wave surfers, Mark Foo was a man who both acknowledged and embraced the deadly potential of big waves, and volumes have been written and hypothesized about his death. Surprisingly, though, little of that story has focused on the guy who had the very last living contact with him. When Mike Parsons was drowning at Maverick's, he assumed he'd been bouncing off Brock Little; only afterward did he realize that Little was nowhere near him, and the person he met trapped on the seafloor was Mark Foo. In the years after Foo's death, Mike Parsons would go on to earn two XXL awards (the Oscars of the big wave set), two big wave world records (both set at the Cortes Bank), and become the subject of the most downloaded surf video on the Internet. Despite this, most folks outside of surfing don't even know who the hell Mike Parsons is. He has never garnered a feature on *60 Minutes* or graced the cover of *Outside, National Geographic, Details, Men's Journal,* or any other mass-market magazine that purports to cover the lives of hard-core adventurers. That's a shame, really. Parsons is surely among the most fanatical big wave hunters the world has ever seen, and to know him is also to gain an insight into the obsession that drives the sport. The complete portrait of this obsession, though, also includes the crew with whom Parsons shared his earliest Cortes Bank encounters: in particular, his former protégé Chris Mauro, former rival Brad Gerlach, and friends "Skindog"Collins, Peter Mel, and Evan Slater. To understand the addiction to big wave surfing is to know who these surfers were before the death of Mark Foo, and who they have become—or *not* become—since.

I use the word *addiction* deliberately. The simplistic view writes off what Parsons and his friends do as somehow optional or voluntary—people assume that what they do is somehow *chosen*. This leads many to conclude that big wave surfers are either outright crazy or utterly selfish and self-destructive. The truth seems to actually be anything but. Mike Parsons and his friends are prisoners. They just don't seem to know it.

One fine day in 1973, Robert Parsons was alarmed to find his seven-year-old son perched on a high cliff near their home above Laguna's Three Arch Bay in Southern California. "What are you doing up there," he asked young Mike.

"I wanna know what it would be like to take off on a wave at Waimea Bay," was the reply.

Bob wasn't terribly surprised. When Mike was three, he knocked out a tooth downhill skateboarding. A few months later, he learned to ride a bike without training wheels. When Mike was four, Bob paddled him into his first waves at San Onofre, and by seven, Mike had won his first contest. "He had such athletic ability and was just so determined from an early age," says his mother, Jodi.

Mike ran, leaped, and fell with seemingly reckless abandon. He excelled at soccer and little league baseball, and when his parents took him skiing, he would find the best guy on the mountain and follow him down like a kamikaze pilot. But Bob and Jodi soon noticed an incongruity in their little hellman. Mike may have been nearly impossible to track, but he was well-behaved and seemed to generally abide by his parents—sometimes to a fault. One day at Big Bear Mountain, Bob sent his tyke up the rope tow with the instructions, "Don't let go." Near the top of the hill, a sharp-eyed ski patroller realized that Mike was moments from having his hands run through the massive pulley at the end of the line. "I said, 'Mike, you'd have gone through that and crushed your hands.'" Bob recalls. "He said, 'Dad, you told me to hang on—so I did.'"

When Mike was six, his family moved to Three Arch Bay. Mike's boyhood home is a small, one-story shake-shingled cottage, built during an era when

Laguna was a far less expensive community of artists, hippies, and cosmically tuned wave riders. It's set just off the Pacific Coast Highway above a thin crescent of desert chaparral, palm trees, blond sand, and azure sea. The waves aren't world class, but they're punchy, fast, and perfectly suited to high-performance wave riding. When the swells get big, the outside peak can be downright intimidating. "It was the ultimate spot to grow up," says Mike. "Almost surreal. It's hard to believe I had such a setup as a kid. It was the kind of place where parents would just let the kids go, and you could just run around all day and night. When the waves were good, there would only be three, four, maybe five of us out. And in the winter, when it got big, I was always the ringleader—even from an early age."

During these years, a glitch was revealed in Mike's auditory processing that made it initially very difficult for him to learn to read. He struggled through elementary school and was heckled even by his friends. He was eventually enrolled in a visual learning program at a school an hour away—a place he hated. Yet he rarely missed a day, and chose instead to channel his anger and embarrassed frustration into bravery and athleticism. "I wanted to do something no one else could," he says. "It was like, people can say what they want about me, but I can prove that I'm better at something, too—I'm braver than they are. I'd skateboard the highest hill, go the fastest, and crash the hardest. When I started surfing, I wanted the wave of the day. I would be pissed off and completely distraught if someone caught a bigger wave than me."

In the ensuing years, Mike would go on to enter arguably more surf contests than any person in history, and he has *never* missed a roll call for a heat. It's also likely that no father ever watched his son compete more. Bob Parsons developed a reputation among Mike's friends as a fervent surf soccer dad. When I ask about it, Jodi points out that Bob was also a fanatical mountaineer and highly competitive volleyballer. "I don't want to talk about it," she says with a laugh. Then she pauses. "Bob was intense. He'd get mad at the judges. If another surfer cut Mike off, he'd get mad. God was he competitive."

Bob admits he could, perhaps, go a bit too far, yet Mike insists that his dad only truly pissed him off once. It was during the 1986 Sunset Beach World Cup—one of the most prestigious events in surfing. Mike was a twenty-year-old pro who had the made the finals alongside powerhouses Hans Hedemann, Mark "Occy" Occhilupo, and Sunny Garcia.

"They were calling the finalists to come up onstage," Mike says. "Me, Hans, and Sunny were there. But Occy hadn't shown up. Dad saw Occy kind of run off around a house. I think he was smoking some weed or something." Parsons laughs heartily and continues, "My dad went up to him and made a comment—something like, 'It was good of you to finally show up.' Occy starts yelling at my dad. Then he comes over to me and says"—Mike uncorks a pitch-perfect imitation of Occhilupo's boyish Australian accent—"'Hey mate, your dad said this and that. I'm going to fight him after the final.' I was like, 'Whatever's going on with you and my dad is between you guys.' He's yelling the whole time. I got third, and I was so mad at my dad that I said, 'You're never going to come to another contest.' My dad was so upset. He went to Occy's house that night— knocked on his door and says, 'I owe you an apology. I want my son to win so bad, I was out of line.' Then Occ was like, hugging him and saying, 'Aww mate, I love my parents, too. They're just as into my surfing.' From then on, he and Occ totally hit it off. He was definitely the dad who wanted me to do well."

"That was probably the stupidest thing I ever did," says Bob. "But Mike's right. I was just so interested in seeing him do well."

By age twelve, Mike was the unofficial leader and drill sergeant of a crew of young Three Arch Bay rippers. A young disciple named Chris Mauro became his most ardent and terrified disciple.

On a day when Chris's parents might think he was at the Parsons's house, he would be in Santa Cruz or Ventura at a contest. "It's 12 feet and I'm this little guy in the junior's division trying to sleep with the gearshift in my back—just scared shitless," Mauro says. "Mike would battle this guy from Ventura named Barry Wilson to see who could get in the water earlier. He'd be like, 'Come on you little fucker, I don't want Barry to beat us.' When we'd go to Trestles, I'd have to show up on the dark front porch of his house at 4 A.M. Freezing my ass off. But if you were five minutes late, you'd miss the bus and you'd be hearing about it. "

Mike's posse held mock contests, pretending to be surfers like Simon Ander-son, Rabbit Bartholomew, and Shaun Tomson. Mike named his dog Shaun. He sent his friends fake letters, saying that Tomson's nascent surf company, Instinct, wanted to sponsor them. At some point, for no apparent reason, Mike's friends started calling him "Parsnips," a moniker eventually shortened to "Snips."

Snips's surf heroes didn't smoke pot or drink—at least so far as he knew—so he concluded that he would live the same way. He became surfing's answer to Richie Cunningham—with his freckled Irish complexion, closely cut red hair, and skinny build, he even looked like a young Ron Howard. "My dad just drilled it into me when I was young—don't take things from strangers, and drugs will take away from your dreams. I don't think I was a total Goody Two-shoes. I mean, I tried smoking weed once or twice, but I was like, 'What is this? This is just going to get in the way of my deal.' I chose my path. I wanted to be the best surfer in the world."

Mike made his first North Shore pilgrimage at thirteen. If his fearlessness scared the hell out of Bob, his judgment left his dad somewhat reassured. Walk-ing from the water after a session at big Sunset, a burly Hawaiian relieved Mike of his board, calling it "an aloha tax." Mike possesses what Bob calls "a high strength-to-weight ratio." A scrap was debated, but Mike wisely recognized

George Hulse, December 1990, on what was long thought to be the first wave paddled into at Cortes Bank with Bill Sharp paddling on the shoulder. "There was definitely this feeling of incredible speed—of how quickly you were moving down the Bank—like moving down a conveyor belt," said Hulse. "I guess because the waves were coming out of the open ocean." Photo: Larry "Flame" Moore/A Frame Photo.

The first known chart and mapping of the Cortes Bank by the United States Coast Survey of 1853 by Lieutenant T. H. Stevens of the USS *Ewing*. The name Cortez was eventually corrected to Cortes, reflecting the first American ship thought to have discovered the Bank. This map does not note the existence of the shallowest reach, today known as Bishop Rock. This rock would be discovered two years later by a U.S. Navy Lieutenant, Archibald MacRae, a hardcore young adventurer who would die a horrible death shortly after setting sail from the Bank. Image scan courtesy of Steve Lawson.

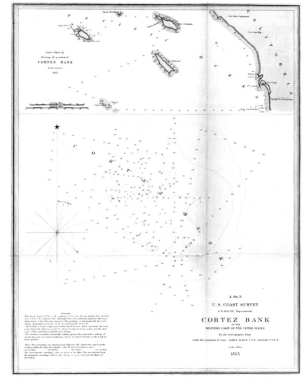

No image of Lieutenant Archibald MacRae, who discovered Bishop Rock, can be located. We might imagine, though, some resemblance to his dashing father, General Alexander MacRae, 1796–1868, a man who served the United States during the War of 1812 and the Confederacy fifty years later. The general would bury at least three of his nine or ten sons (it's not entirely clear how many he had)— with Archibald being the first. The general's letters show a man both fond and proud of his explorer son. In 1840, he wrote: "I am afraid that you do not take the interest in writing home that you aught. 'Be not weary in well doing.' Don't neglect in writing as it is is not only a source of gratification to 'all hands at home'... tell us all that you have seen — that is, all that is proper to be written. Your affectionate father, Alex MacRae." Photo courtesy New Hanover Public Library, St. Johns Masonic Lodge No. 1 Collection.

James Whitemarsh, Executive Officer of USS *Ramapo*, veteran of World War I, a future veteran of Pearl Harbor and Iwo Jima, and witness to the largest wave ever recorded from the deck of a ship in 1933, marries Rebecca Bird Caldwell Gumbes, future mother of Francis "Taffy" Wells. Photo courtesy of Whitemarsh's great niece, Angie Gregos-Swaroop.

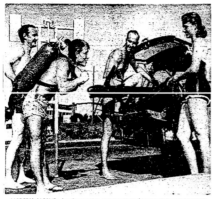

MODERN JASONS—Loading station wagon with underwater diving equipment in preparation for treasure hunt 100 miles southwest of Palos Verdes Peninsula are Richard Ormsby, Richard Stoeffel, Mel Fisher and Dolores Fisher, left to right.

TREASURE MAP—Chart shows Bishop's Rock in Cortes Bank, area where Spanish galleon with $700,000 in gold and silver reportedly sank in 1717. Exact spot is just below figure of Neptune on map. It is almost due west of San Diego.
Times photos by Hubert McClain

UNDERSEA CAMERAS—Mel Fisher, left, holds camera which will take still color pictures under water, while Richard Stoeffel holds 16-mm. movie camera.

WOOD SEA SLED—Making mock-up for undersea sled which will carry electronic metal locator are, from left, Richard Stoeffel, Mr. and Mrs. Mel Fisher.

DREDGED FROM DEEP—Mel Fisher displays bronze mounting from old sailing ship and bronze spikes he took from hull of sunken ship explored last year.

DIVERS WILL HUNT UNDERSEA FORTUNE

BY LEE BASTAJIAN

A sunken Spanish galleon, holding $700,000 in gold and silver, is the object of an adventurous group of Southern Communities skin divers·

The ship's hull lies in 75 feet of water beneath Bishop's Rock, a treacherous underwater promontory located in Cortes Bank, 100 miles southwest of Palos Verdes Peninsula.

Two Others Follow

The galleon, bound from Acapulco, hit the rock, foundered and sank in 1717. It was joined in 1855 by the Stillwell S. Bishop, carrying a cargo of guns, and the SS El Capitan in 1952, laden with general cargo.

But the wreck of the galleon is the major lure for the group, which includes a trio of Torrance residents, Mel Fisher, underwater specialist and director of the expedition; Richard Ormsby, an airline clerk, and Richard Chap-

man, roller-skating instructor; a Lomita carpenter, Dick Stoeffel, and a Long Beach physician, Dr. Nelson Mathison.

Waiting for Right Time

Toying with the idea of joining the expedition is Fisher's wife Dolores, herself an expert skin diver.

They'll have the newest electronic metal-locating devices, gadgets never before used in undersea exploration, and they'll shove off on the expedition when favorable diving conditions develop at Cortes Bank.

"Normally 50-foot waves break over the rock," said Fisher. "Thus our departure will await a relative calm— 20-foot waves—at the banks. We'll receive daily wind reports for two weeks from the Coast Guard Station on San Nicolas Island. The velocity must be less than 10 knots in order to insure calm water.

FINNY BIKE—Ocean pedaler enabling diver to triple underwater speed is ridden here by Dolores Fisher, and held by Richard Ormsby, left, and Mel Fisher.

Lee Bastajian's *Los Angeles Times* piece on Mel Fisher's first major expedition out to the Cortes Bank in 1956. "The whole thing," *Times* reporter George Beronius would later say, "it was just a complete fiasco." Image courtesy of the *Los Angeles Times*.

Mel Fisher discovered the wreck of the *Nuestra Señora de Atocha* on July 20, 1985. It was the greatest treasure ever discovered on the sea floor. Coincidentally, singing storyteller Jimmy Buffett, a friend of Fisher's, happened to be nearby. "I was fishing about a mile away when the wreck was discovered," Buffett recalled. "Then they called me on the radio and said, 'You might want to come over here.'" In the photo, Buffett is singing "A Pirate Looks at Forty" on the day of discovery, while seated alongside Fisher atop a pile of gold and silver bars from the *Atocha*. Photo and story courtesy of Jimmy Buffett.

A bat ray soars through the teeming kelp forest atop the Bishop Rock — directly beneath the surf zone. Photo: Terry Maas

Another buddy of Mel Fisher's and the first person known to have surfed the Cortes Bank is Harrison Ealey of Oceanside, California. Ealey reports that he accomplished the feat during the summer of 1962 on a big south swell. Here, Ealey holds up a photo of himself dropping in just inside of Buzzy Trent at Waimea Bay, Hawaii, in 1963. "If I'd never surfed in Hawaii, I would have been scared to death out there," Ealey said. Photo: Chris Dixon.

In 1952, Joe "Palooka" Kirkwood Jr., one of the originators of the idea of sinking the SS *Jalisco* atop Cortes Bank's Bishop Rock to create the nation of Abalonia, was at the top of his game. He was a well-paid actor and Masters-caliber golfer who had been invited to play in the Azalea Open golf tournament, in Wilmington, North Carolina. Upon learning that Kirkwood's wife, Cathy Downs, was in town, organizers of the concurrent Azalea Festival offered Downs the annual Azalea Queen crown. In a remarkable coincidence, one of North Carolina's most prolific and renowned photographers, Hugh MacRae Morton, captured Kirkwood with his arms around Downs during coronation of the festival's Azalea Princess (princess's identity unknown). MacRae Morton was the great-great nephew of Bishop Rock's discoverer, Archibald MacRae. Photo: The North Carolina Collection at the University of North Carolina.

The final moments of the *Jalisco* and the nation of Abalonia. One hundred and eleven years and 361 days after the death of Archibald MacRae. The photo shows Joe Kirkwood Jr. perched out on the bow, the instant before he was blown off his great ship of state. Standing in the lee of the three-story-tall superstructure, Jim Houtz recalled Kirkwood's last words: "The wave's gonna go by me. It's gonna wash around me."

Taken shortly after Joe Kirkwod's last moments aboard his ship, this photo shows a massive wave breaking over the *Jalisco* in the take-off spot today preferred by surfers like Mike Parsons and Greg Long. The ship's jagged, battered hull still rests below the surf break. The photo also shows that the ship's three-story-tall superstructure has been obliterated. Both photos: Associated Press.

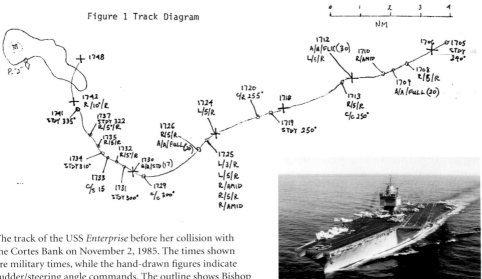

Figure 1 Track Diagram

The track of the USS *Enterprise* before her collision with the Cortes Bank on November 2, 1985. The times shown are military times, while the hand-drawn figures indicate rudder/steering angle commands. The outline shows Bishop Rock's six fathom (maximum thirty-six-foot deep) outline. The *Enterprise* passed through very shallow water, barely two nautical miles from where the waves break. Image courtesy of Karlene H. Roberts taken from the study: "Bishop Rock Dead Ahead: The Grounding of USS *Enterprise*." By Karlene H. Roberts, University of California, Berkeley. 1986. Image courtesy: Karlene H. Roberts.

The USS *Enterprise* in 1966. Photo: U.S. Navy.

WAVE HEIGHTS / SWELL PERIODS / ENERGY DEPTHS WITH 40-KNOTS SUSTAINED WIND

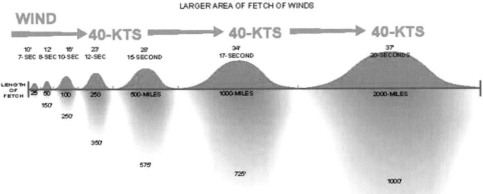

This image shows the effect of a constant forty-knot wind on the surface of the ocean. After one hundred miles, a swell is fifteen feet tall and two hundred feet deep with a ten-second period from trough to crest. After two thousand miles, it's thirty-seven feet tall and one thousand feet deep and carries a twenty-second period—and orders of magnitude more power. A swell like this could create a perfect breaking wave more than a hundred feet tall atop the Cortes Bank. Image courtesy: Sean Collins/Surfline.

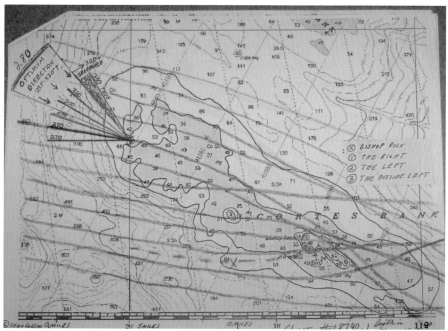

Sean Collins's slavish devotion to waves is shown in his calendar from 1986, just one of many pages of a years-long documentation of every swell to hit Southern California. Below that, one of Collins's early hand-drawn charts of Bishop Rock, showing his best estimation of prime swell directions and periods for Cortes Bank. Collins keeps the very best angles and periods a secret.

Peter Mel and Ken "Skindog" Collins. Project Neptune, January 2001. "It was awesome," said Mel. "Heaven on earth." Photo: Rob Brown.

Mike Parsons's first wave at Cortes Bank on January 19, 2001, would also become his first world record. Note Evan Slater and twenty-one-year-old Captain John Walla sitting thunderstruck on the shoulder. They had both nearly died a few minutes earlier while trying to paddle in. Photo: Aaron Chang.

Another view of Parsons's epic ride. "The size and the sound—it was just such a leap over anything I'd ever seen," said Rob Brown. "I was sitting there going click, click, click, watching the counter go down frame by frame, telling myself, 'Relax. Mike is gonna die right here and right now, but you're gonna do your job.'" Photo: Rob Brown.

Parsons's 2001 wave as seen from the air by Bill Sharp, Vince Natali, and Larry "Flame" Moore. Photo: Larry "Flame" Moore.

Evan Slater (left) and Captain John Walla (right) are all smiles after nearly dying trying to paddle in at the Bank during the Project Neptune mission. On December 23, 1994, Slater and Mike Parsons pulled the lifeless body of Mark Foo from the water at Mavericks. Photo: Rob Brown.

Brad Gerlach, perfect form, Cortes Bank, January 19, 2001. "I wish I was there right now. I think about it so much. I wish I was there." Photo: Rob Brown.

Bill Sharp, mapping out *The Odyssey* sometime in the early 2000s. Photo: Les Walker.

Mike Parsons's second ever wave at Jaws, January 7, 2002. A still photo of the most downloaded surf clip on YouTube and the poster shot for the film *Billabong Odyssey*. Helicopter pilot Don Shearer and motion picture cameraman Peter Fuszard look on from the bird's eye position. Photo: David Pu'u.

Mike Parsons and Brad Gerlach contemplate their great white whale on a raw winter's day in 2003. Photo: Grant Ellis.

Greg Long, December 17, 2003, on a wave that would land him his first *Surfer* magazine cover. "The wave of my life. No questions asked." Photo: Rob Brown.

Greg Long and Grant "Twiggy" Baker stare into the abyss at the Cortes Bank on the Everest expedition, January 5, 2008. The wave ahead of them is easily higher than 80 feet. "As far as the eye could see, it was just a huge square of whitewater," said Twiggy. "If you lost your guy in there, he was just gone. He would have been lost in that expanse, and you'd never find him. It was just so scary." Photo: Rob Brown.

A diminutive Mike Parsons on what today remains the largest ride ever documented, on January 5, 2008. This wave has been estimated to be around 80 feet high. The height of the exploding whitewater is anyone's guess. Photo: Rob Brown.

A quiver of paddle-surfing boards on the deck of a boat that would set sail for Cortes Bank, on December 26, 2009, carrying a team of the best paddle surfers on earth. The boat would later be nearly swamped by a rogue wave atop the Bank. Photo: Jason Murray.

Greg Long paddles into a cerulean monster and defines the future on November 10, 2011, atop Bishop Rock. "When you're paddling all alone out there, when you really look at the place and feel its immensity," he says, "you can't just help but feel that there's something so much greater—so much more significant at work than you." Photo: Chris Dixon.

Three generations of Cortes Bank fanatics. From left: Mike Parsons (world record big wave surfer), author Chris Dixon, Jim Houtz (a former world record setting deep diver and a founding father of the nation of Abalonia), and Greg Long, back in San Clemente after a successful if not terrifying tow and paddle-surfing mission to the Cortes Bank, November 10, 2010. Photo: Chris Dixon.

Irascible, pioneering, devoted, faithful. This book would not have been possible without the work of the late, great Larry "Flame" Moore. Photo: A Frame Photo.

he might invite a world more trouble. "Mike was daring," says Bob, "but he wasn't foolish."

Bob financed Mike's second competitive journey to Hawaii with a second mortgage on the house. This was when Mike's surfing began to terrify Jodi. "The first time I went to Pipeline, here comes Mike walking up the beach with his surfboard in two parts," she says. "I asked him, 'If it did that to the surfboard, what did it do you to you?' He just said, 'I'm fine, Mom.'"

Hawaii increased Mike's bravery fivefold, and Chris Mauro suffered for it. One day Mike egged the grommet to the outside at Three Arch Bay on a massive wintertime swell. Petrified over paddling in through the slabbing shorebreak, Mauro started crying. One of the surfers, maybe even Mike, called him 'sis' for sissy. The name stuck. "Then I felt really bad for him," says Mike. "He was my little right-hand man. I'd tell him what to do and he'd do it. Chris just had so much talent. He was a real natural. I wanted to encourage him like my dad did for me. So I basically put him on my shoulders and said, 'If you want this surfing thing, you can do it. I paid for his meals—every meal he ate—and even entered him in contests when he couldn't afford it."

Chris Mauro eventually followed Parsons to Hawaii with the National Scholastic Surfing Association's (NSSA) National Team. Parsons would order the bug-eyed grom out in the predawn darkness before either surfer even knew how big it was. He loaned Mauro brand-new boards and ferocious waves broke two in one day. Mauro also worked as Mike's board caddy during the World Cup contest at Sunset Beach—ferrying out a new stick when Mike snapped one. "I was this little kid," Mauro says. "I'd take out a board and have to *swim* in— and it was *massive*. When Randy Rarick, the contest organizer, spotted me, he goes, 'That little kid can't be your caddy!' We stayed with Peter McGonagle one time—this hard-core San Clemente guy. Back then an 8-foot board was considered a big rhino chaser. Peter had this 10-foot 6-inch gun. I was like, 'Holy shit, what is *that* for?' He looked at me like I was crazy and said, 'Dude, it's for *outer reefs*.' Those are the guys Mike surrounded himself with. It was heavy."

It was around this time that a cocky, witty, handsome, and supremely talented kid named Brad Gerlach joined the National Team. Mike and Brad soon developed an interpersonal rivalry that would define their surfing careers. If you had told Mike that this brash, noisy punk would one day become his best friend and guardian angel, he would have laughed out loud.

In some ways, Brad Gerlach's (pronounced *gerlock*) Aquaman future was preordained. Gerlach was born in 1965 to a stunning water ballerina, while his father, Joe, was a Hungarian high diver who had finished fourth in the

1956 Olympics. After defecting to the United States in the wake of a bloody Hungarian revolt, "Jumpin' Joe" Gerlach earned a pair of collegiate national championships. He then revolutionized the sport of pole vaulting by inventing a thick series of pads about the size of a Chevy Nova for jumpers to land on. Eventually, he dove into a livelihood by performing spectacular leaps onto these pads from terrifying heights. On the day before he was to be wed, a slight aerial miscalculation shattered his facial bones like an eggshell. "I look much better with this beard," Joe says today with a laugh.

One of Joe's biggest leaps of faith came in 1971 during halftime of a Lions-Eagles football game, an event described thusly by *Sports Illustrated* writer Sandy Treadway:

> A Sunday night national television audience estimated at 30 million saw a giant red, white, and blue balloon ascend 80 feet above the 50-yard line. A slight man dressed in a white leather jump suit stood precariously on a wooden plank attached to the gondola and directed the placement of a foam rubber mattress—his landing pad. After several minutes of tension-building waving, the man appeared satisfied that the sponge was properly aligned and, at last, he raised his arms and stood on his toes. Helmetless, he left the plank in a swan dive and fell toward the field, landing on his back directly on target. He lay motionless for a moment, "checking myself out, making sure I'm alive."
>
> Among those watching on television were Joe's wife Cheryl and their 5-year-old son Bradley, who asked, 'Is Daddy dead?'"

To Bradley, Dad was a true Superman, but he was also gone a lot. If Joe landed, say, a marquee job in Vegas, the family sometimes went along, yet by the time Brad turned seven, the long absences and travel took their toll, and Joe and Cheryl split. Brad became a troublemaking latchkey kid in an Encinitas duplex. He sees some parallels of his early life to that of Laird Hamilton, whose adoptive dad, Bill Hamilton, was a larger-than-life surf hero also AWOL during stretches of his childhood. Brad says he felt "definitely a lot of pride in my dad. But also a lot of anger that he wasn't around and wasn't paying me much attention. I know Laird got in a lot of fights when he was younger. He probably could have used a little more attention from his dad, too."

Three years later, Brad found an orphaned surfboard on the beach with a big picture of Jesus laminated beneath the fiberglass. He quickly learned to walk on water and was soon emulating the moves of local pro John Glomb. One day Brad paddled over to a budding amateur, who was his elder by a couple of years. "Hey, hey, hey, are you sponsored by those guys?" he asked, pointing at the logos on Jeff Novak's board. "Umm, what do they give you? Hey, you can come down and surf and stay at my house if you want to. My mom's always gone."

Novak was bemused that this brash little ten-year-old kid would paddle over to a complete stranger without fear of being insulted—or punched. "I'm thinking who is this kid?" Novak laughs.

Before long, Jeff Novak and buddy Ted Robinson were hitching rides down to Encinitas to stay with Brad. Cheryl worked long hours, but when she was around, there was a lot of laughter. "She was just, just *stunningly* beautiful," Novak says. "You know, back then we're kids, and his mom was *hot*. She was so cool and just really enjoyed talking to Brad and his friends."

Brad fed off Novak's experience, but awe at his dad's wildly unorthodox career mingled with resentment over his absences also pushed him to compete. "That anger fueled my drive," Brad says. "I wanted to be the best guy surfing."

Novak and Brad pushed one another hard, and Brad mapped out his plan by age thirteen: compete in the NSSA, then the Professional Surfing Association of America (PSAA)—a sort of stepping stone to the big leagues—and then graduate to the Association of Surfing Professionals (ASP) World Tour. His obsession bore early fruit when he was signed to John Glomb's Nectar surfboards team at age fourteen—a heavy accomplishment.

A few years earlier, in the winter of 1976 when Brad was eleven, Joe Gerlach shared a Chicago bill with Evel Knievel—a show neither would perform. On the same day that Joe's plane crashed into a snow bank on its final approach to Chicago, Knievel crashed during a practice attempt to jump over a tank of live sharks, breaking both arms and taking out the eye of an ABC cameraman. Both hellmen took the day as a sign and quit jumping forever. Knieval retired rich. Joe bought an inflatable dome and a machine that projected a mechanized laser light show to Pink Floyd songs and took the show on the road.

After that, Brad began to travel with his dad during the summers, working as a carny. He visited towns in New Jersey, Pennsylvania, and Arizona that few Southern California kids ever see. He smoked joints, drank beer, and performed grand swan dives into huge piles of dad's beanbag jump cushions. When he wasn't with his dad, Brad and Novak tooled up and down the California coast, with a far-too-young Gerlach tumbling out of Jeff's VW camper beneath a halo of smoke. "But no matter how hard we partied, we were always up first thing the next morning surfing," says Novak.

Brad increasingly found nothing but trouble in Encinitas and moved to Huntington Beach with Joe when he was fifteen. "Boy, he used to get in fights all the time," says Joe. "One time he took my Leica camera down to the beach and these guys wanted to take it from him. He later said to me, 'I just can't imagine what you would have said if I'd lost your camera.' So he went to do this jiujitsu kind of thing in front of them, and they backed off. I mean, he just *made that shit up.*"

A few months later, Brad came home one afternoon particularly frustrated after losing a revered competition called the Katin to guys he *knew* he could outsurf. "I'm gonna fucking quit," he told his dad.

Joe didn't know jack about surfing, but he possessed an Olympian's athletic instincts. He grabbed a pair of binoculars and joined Brad on the beach. "We had so many funny arguments," says Brad. "He'd be like, 'Your butt's all over the place.' I'm like, 'You don't even fuckin' surf!' Then we got a video camera, and I was like, 'Oh, I see what you're saying.' He was hands-on, watching me surf, where I was breaking my line, telling me not to turn so hard when you don't have enough speed. He wanted my surfing to be relaxed and have a beautiful look. Surfing's an aesthetic. If you look good, you score well. We watched ice skating, gymnastics, diving. He'd go to the beach with me every day, twice a day. He'd come to every contest. We'd talk about me and we'd talk about surfing. I really needed that because I was really learning about myself. He was the only one who wanted to talk to me more about my surfing than I did. He became my best friend."

When Gerlach was inevitably invited to join the NSSA National Team, he and Snips were arguably the two most driven and talented guys on the roster. They were also the most different.

"I'd just see him and be like *Parsons*," Gerlach hisses. "Like this guy's really going to beat me. He was the model team rider, tucked his sweatpants into his Ugg boots, and he'd have his wetsuit on and there's just not a drop of water out of place.

"He'd look over at me in exactly the same way. '*Gerlach.* What a disheveled. . . Probably didn't even come down to the beach with wax on his board. There's no way I'm letting this guy beat me.'"

Parsons agrees, musing, "There was *no one* I wanted to beat more."

Joe wanted Brad's coaches, Ian Cairns and Peter Townend, to take a more technical approach, and he didn't mind sharing with Brad that he disagreed with their methods. This stoked Brad into defiance that landed him regular NSSA suspensions. Temperamentally, father and son were not so far apart. "I was very arrogant," admits Joe.

Brad turned pro at nineteen, followed shortly by Parsons—each was paid a thousand dollars a month by the nascent Association of Surfing Professionals. Surf magazines of the day hyped duals between Gary "Kong" Elkerton, Tommy Curren, and Mark Occhilupo, but they didn't pick up on Mike and Brad's rivalry. Indeed, neither surfer even realized how much one despised the other till much later on. Yet Mike and Brad both feel that their quiet rivalry was actually the most intense—because it revolved around a validation of their very identities.

Parsons first served notice in Australia, taking second to eventual world champ Curren at Burleigh Heads. Then, in one of the biggest upsets in history to that time, Gerlach defeated reigning world champion Tom Carroll at the Stubbies Pro in Oceanside, California. *Surfer*'s prophetic scribe Derek Hynd wrote: "In that instant, watching a young surfer in form, words from old British punks came forth: the next generation; a revelation."

Gerlach ate up the hype and gave hilarious interviews in return. Here is a typical response in a *Surfer* profile.

What would a dream date with Brad Gerlach be?

Well, I think it would start out with just driving somewhere really fast. Me and my date, in my new car. As fast as it could go, like 140 or 150. Headed for . . . I don't know . . . Headed for the desert. To some kind of small nightclub, nothing like hip or trendy. And there's all kinds of people there, warm people—people who don't know me. . . . It's just really friendly, it's totally happening, and as soon as we get out of the car, the first person we see yells out, "Alright! How you doing?" And from that point on, we're on a roll.

. . . And then—shit, I don't know. It's hard for me to plan these things, they just happen. No, okay, then we'd meet someone there, at the club, and they'd suggest something to do and a group of us we would be all, "Yeah! Let's do it!" And we're off. Like we'd go slide down a hill on a big piece of cardboard or something.

So we're doing this and we're doing that. . . . It's constant movement. We'd drive to Vegas. We'd gamble a bit and win a lot of money. We're on a roll. We'd get married, then shine it on a couple hours later. Just for a joke.

Then we'd buy skis and equipment and go to Aspen for a couple days, staying in a cabin up on a hill. And the only thing in the living room is a fireplace and a killer stereo. Like a $10,000 stereo system. Music's key. And then we'd fly to Key West to get a tan.

My dream date could actually go on for months.

Gerlach laughs and recalls something he recently told Chris Mauro. "If I met me at nineteen today, I would *kick my ass*."

During their days in the ASP, Gerlach and Jeff Novak reveled in their exotic travels and were baffled at the apparent lack of interest in foreign cultures displayed by some of their peers—notably Parsons. Mike pleads guilty. He and Mauro could be driving through the most bucolic slice of French countryside, but his head was buried in a heat sheet. "I could be in a country where I'd never even been before, and I barely even noticed I was there," Parsons says. "All I noticed was the beach and the waves and how to win my heat. I had that crazy, like, single focus."

You can see Parsons's nature at work still today—all you have to do is paddle out with him. Not too long ago, I marveled as he surfed at Upper Trestles with his best friend Pat O'Connell, a star of Bruce Brown's film *Endless Summer II*. Not only were they dissecting the velvety righthanders, but they seemed to be holding an impromptu contest—critiquing and scoring one another's rides. Parsons admits, "That's totally what we were doing. I don't want my friend to outsurf me—it's always competitive. I know that sounds weird because sometimes you're totally enjoying yourself—and sometimes, well, sometimes you're kind of not. Especially on some of these big wave deals. But when you're around your friends and guys who surf well, you don't want to be a kook or lose your edge. If you feel like you're slipping, even in little waves, it's the worst feeling. That's kind of my biggest fear in life—not being able to surf at the highest level. That's just a horrible thought."

In this way, Gerlach and Parsons were true opposites. "I always wanted to win—bad," Gerlach says. "But Jeff and I also wanted to experience France, for example—to drink wine and eat croissants, try the cheeses, and learn the language. We didn't go to college, so we were out there just getting educated on, say, Bastille Day. What the hell is Bastille Day? Well, it's obviously a big deal. They're blowing shit up and all the French girls are happy. What's to all this? I wanted to stay with French families—study the way they went about eating. It's such a great way to get to know people—eating, sharing—breaking bread. I'd stay up with families who barely spoke English telling stories, and it would mostly be charades. We just had amazing times."

While Gerlach was making everyone laugh with charades, Parsons developed an ability to disarm his competitors with charm while plunging in the competitive knife. He was so damned nice, and possessed such a savant's devotion to surfing, that he became one of the most popular guys on the tour. When an Australian surf magazine dared question whether Parsons was a future world champion, the normally nationalistic Aussie pros threatened the writer's life. "They were like, 'Fuck you, dude'—totally protective of him," Mauro says. "That's in part because Mike was never like Ken Bradshaw, Mark Foo, Brad Gerlach, or any of these guys who promoted themselves. He didn't and doesn't claim anything. He earns respect by competing and being out there every day. If he sees someone already getting back to their car after surfing and it's 6:30 in the morning, that to Mike is the best surfer in the world—because he's the most committed."

Yet it was that commitment to surf and nothing else that eventually drove Mauro and Parsons apart. Early in what Mauro calls his "flash-in-the-pan surfing career," he defeated Parsons in Fiji. There, he began to see his longtime mentor and drill sergeant as an equal. He bristled when Parsons treated him like a little brother, and he quickly tired of the endless routine of airplanes, cheap hotels, and nonstop competitions that fueled Parsons's fire. "He'd be like, 'I'm driving the rental car,'" Mauro says. "But I'd paid as much for the car as him. I wasn't his sidekick anymore. I've analyzed our thing, and he and I laugh about it today. He was so singularly obsessed with surfing—getting his fix—and he's still that way. Mike hitches himself to the people who are as hooked as he is. He's obsessive-compulsive, and surfing is his biggest vice."

Mauro found his limits, and happily so, both as a competitive athlete and a big wave surfer. He was and has remained a committed surfer, but he reached a point where he wanted to move on, maybe take a stab at writing (which would eventually lead to an editorship at *Surfer*). This, of course, led to the end of the "Sis" and "Snips" surfing partnership, but they remain close friends.

Yet really, Parsons's focus was shifting, too. In a lengthy 1989 *Surfer* profile, Mike admitted that competition wasn't "everything." If I find myself getting disappointed with all the travel and contests—which I definitely do now and then—something right always seems to happen. Like, I'll take three days off in South Africa and get 8-foot Jeffreys Bay. You get a fix like that, you know you're on the right track—in life, I mean. And that's not a competitive thing."

Parsons had climbed well into the Top 16 on the ASP World Tour, but rarely actually won a contest. In the ASP's more typical small wave competitions, he was too calculating, too precise. He performed best in big wave contests, where he didn't have to think too much. Eventually, eleven months a year on the road took such a toll that even the competitive machine was worn down. Parsons wanted to focus on surfing big waves and competing closer to home. He joined the U.S.-based PSAA Bud Tour in 1991—a contest circuit ironically held at mostly small wave venues—and won the whole damn thing.

The same year, Brad Gerlach finished second in the world to Aussie Damien Hardman. Many felt Gerlach was robbed. Gerlach had been featured on a few MTV shows—where his looks and hilarious, chatty style suited the network perfectly. He wanted to set off in other direction—film, music, a move to LA. Jeff Novak thought he was crazy. "We took a long walk," Novak recalls. "I said, Brad, it's great you have those aspirations, but you're second *in the world*. I did all I could to talk him out of it, but he was like, 'No, I've already made up my mind.'"

Yet stepping off the tour was a mixed bag. Musical success didn't come immediately, and Gerlach found that he actually didn't *like* the celebrity and loss of anonymity wrought by MTV. He stepped out of the spotlight, choosing

ing a low-budget life chasing waves and photo incentives around the world. It was a pretty decent life, but he was no longer a rock star. Brad and his father, Joe, founded an entirely new type of team-based surf competition, which they called "The Game," and Joe invented a revolutionary downhill skateboard called a Carveboard, whose turns greatly mimic those on a surfboard. Neither effort made either one of them rich.

Joe is in his seventies today, and I once asked him if he looks back and sees much of himself in his son. He says it's something he thinks about often. "At first, I'd only notice [similarities] when I'd see the bad stuff," he says. "When I want to advise him on something and he doesn't listen. But then, sure, he surprised me sometimes. I mean, I used to think I was wasting time with the life I chose—my friends were in jobs and VP executive positions, building up collateral in their jobs, and what am I doing? But then I realized, I learned more than all those guys put together. I mean, I had to do all this shit myself. I negotiated, I traveled, all this stuff. It was my way to keep my freedom—from having to have a real job. Then Brad—he ends up doing the same kind of radical stuff as me, and I never *told* him to do any of those things. You go for broke, dude. I've done it all my life. So has Brad."

Through the 1990s, though they didn't realize it, Gerlach and Parsons were on a collision course. They were not what you'd call friends after they stepped off the tour, but they were making the same choices: preferring the quality of their surfing life over competitive glory, world titles, and celebrity. They occasionally crossed paths in the water, particularly at Todos Santos, but were initially nothing more than cordial. Gerlach loved to surf Todos, but he also possessed an appreciation for his own mortality that Parsons and a few of his friends—like Evan Slater and a few of the boys from Maverick's—seemed to lack.

"I was in awe of Mike," says photographer Rob Brown. "At the same time, he was just scaring the hell out of me. He would absolutely, 100 percent take off on anything. He didn't say, 'I'm going to charge it,' he just did it. One time I taped an expensive water housing to the nose of his board. He paddled out and took off on a 30-foot closeout—got absolutely murdered—and we never saw the camera again. I was like, 'Mike, what the hell?' He just said, 'I'm so sorry, but God, I just had to go.' He just couldn't pass *anything* up. It was bizarre. There was a period where I was seriously worried that he was losing his mind."

"Rob's observation is pretty good," Parsons says. "I'd watch footage of guys at Waimea, and they'd always come up. At Todos, there was a series of winters where I felt I could take *any* wave and survive it. I'd take off from a place where I knew I wouldn't make the wave because I just wanted to see what would happen.

One time, I sat way out for a couple of hours and waited till the next really big set came through. I took off and got to the bottom of the wave. The lip landed on me and I put my knee right through my board—a horrendous wipeout. I don't think that even slowed me down. I went even harder after that."

One day around the turn of the twenty-first century, no one seems to remember exactly when, Parsons was out towsurfing small- to medium-size Todos Santos with his buddy Taylor Knox. A *panga* showed up bearing Brad Gerlach. Gerlach was curious as hell about the ski, and eventually Parsons offered his former enemy a turn with the rope. Gerlach was suddenly making big, swooping carves—the sort of gouges he might have made at 6-foot Rincon, but at *Todos*. Holy shit, this was fun. Maybe even addictive.

Gerlach had never really spent a lot of time talking to Parsons. He always thought Parsons was so serious. Instead, it quickly became evident to the former enemies that they had a hilarious, adventure-filled past to look back on and laugh about. In fact, Parsons made Gerlach laugh like hell. They started recalling contests, paddling competitions during the NSSA years around the Huntington Pier, when Gerlach was so hungover he could barely see straight, but *God* if he let *Parsons* beat him. The time Gerlach convinced a buddy to sound a contest horn a second before Parsons caught a wave. There were so many moments that each wanted to kill the other, but said nothing. If only they had looked past the surface—what each thought the other represented at the age of nineteen—they might have found a perfect yin-yang balance. Had they been there to critique and encourage, instead of willfully disdain and misunderstand one another, each might have eventually walked away with a world title.

One thing was also obvious. They were complete fiends for this new sport, and they suddenly became perennially bitchy and inseparable best friends. "When we started to hang out before we surfed Cortes, I was just blown away at how much of a nutcase, what an addict Mike is for big waves," says Gerlach.

Parsons will readily admit it. Up until the death of Mark Foo, he didn't really think about dying in big waves. Didn't admit that it could happen to him. It wasn't really until he actually saw Foo lying in his open casket that the reality of what had happened crashed down around him. "That I was there firsthand to have that experience. I was right down there underwater with him in my arms," he says. "Just this realization, you can just . . . *drown*. Everything was running through my head for weeks on end. I *still* have crazy dreams today, when I think back. What could I have done differently? How did that happen?"

Afterward, Parsons paddled out with Ken Bradshaw at 6-foot Sunset Beach—the kind of conditions he would have previously laughed off—and was utterly terrified. He wondered if he'd ever manage to muster the nerve to paddle out at Todos or Maverick's again.

Thinking back today, he furrows his brow and takes a deep breath. "You know, I suppose in the back of my mind, I guess I always knew something like that could happen. But going through it so vividly. Riding the wave behind Foo and bumping into him underwater. All that. It was as hard as you could ever lay it on. I guess that sounds kind of weird. I mean it's nothing—it's *nothing* compared to the fact that he drowned. I mean, Mark lost his life, so who cares what happened to me. But at the same time, it was so *real*. You can read about a guy freezing to death on Mount Everest and you could say, 'Wow, that's so heavy. I can't believe people could walk by this guy, and he's freezing to death in a little cave.' But if you're the guy who sits down next to him and almost freezes to death yourself, it's obviously a lot different. What if those two waves took out three people instead? Because it was that close."

After countless hours of conversation with his dad and friends about the loss and the fear, Parsons came to understand that even if he had realized that he was bumping against Foo underwater, he could have done little to save him. There were no rescue crews out on Jet Skis, and even if Parsons had somehow managed to paddle back and alert everyone else, Foo would have already been down for too long.

For a year or so, Parsons even thought about giving up big waves. "Ian Cairns—he had a horrendous wipeout at the Smirnoff event at Waimea years ago. Injured his neck forever in that one wipeout. He was one of the few guys who admitted, 'That changed my perspective and my entire life. I never wanted to drown. I'm not going to risk my life like that again.'"

Then, on December 23, 1995, a year *to the day* after Mark Foo died in California, a talented Californian named Donnie Solomon drowned in huge surf at Waimea Bay. It was the first time a surfer of repute had died at Waimea since Dickie Cross's disappearance in 1943. Then on February 13, 1997, well-liked North Shore big wave surfer Todd Chesser got caught inside a massive wave and drowned at an offshore spot on Oahu called Outside Alligators.

Bill Sharp said it was as though a fifty-year run in the casino had suddenly turned up three sets of snake eyes. If Mark Foo, Donnie Solomon, and Todd Chesser could drown, any big wave surfer could check out at any time. The entire sport went through a mortality gut check.

But Parsons didn't quit. Unlike Ian Cairns, the very real possibility of dying wasn't enough to scare him off, and unlike Greg Noll, he hadn't yet found a wave that pulled the monkey off his back. However, he promised himself that he *would* do things differently: There would be no more just leaping off a boat and paddling out into the unknown without studying a spot. Where were guys lining up? What were the currents? What were the escape routes? What were his lifesaving backups? If he had none, he might actually be content *not* to surf. At least that's what he convinced himself.

In the end, despite their claims to the contrary, I would argue that the deaths of Foo, Solomon, and Chesser didn't, and maybe *couldn't*, fundamentally change surfers like Mike, Brad, Peter Mel, Skindog, and Evan Slater. They *would* become a bit more careful, but during the ensuing decade, they would become, paradoxically, even more determined, more dangerous, and more addicted.

Of course, *addicted* is a word fraught with peril when discussing something like *surfing*, which would seem to most to be an entirely optional undertaking. But in discussing Mike Parsons and his desire for big waves, Rob Brown, Chris Mauro, and Brad Gerlach all used the word. More than one big wave surfer has merited the term, and it gets at the distinction between *wanting* to paddle out versus literally *needing* to. However, I wondered, could Mike Parsons really be *addicted* to surfing big waves? Was such an addiction even possible? Rob Brown, who has covered big wave surfing for much of his life, assured me that it was.

In fact, he says, it's the only explanation. Over time, he says, "You come to realize—big wave surfing—it's not about size or muscles. It's really in your mind. I mean, of course you have to be fit, but growing up, I didn't have the ability to surf those waves, and I didn't want it. The ability to sacrifice and be near death every time—these guys are so comfortable with it and they're so addicted that they just don't realize *how radical it is*. I've photographed Indy car racers—Ari Luyendyk and Roberto Guerrero. They would risk locking wheels at 240 miles an hour—but they'd never go out in two-foot waves. And I'd feel safer in an Indy car than in these big waves. I mean, when you crash, the car stops—and you can have a doctor pull you out. You could drive straight into a concrete building and survive. At Todos or Maverick's the crash is just the *beginning* of your problems. The building *is* going to fall on you. You're going to get hit a number of times, and you're all alone. I guarantee you, I could not survive a single wave. Wipe out and it would be instant death by either drowning or heart attack. Even if it's a near-death experience—guys like Mike, they just laugh it off. They just paddle back out. It's really weird. I'm not sure they even get it. The bottom line is that on Earth, I don't think there's a more dangerous thing you can do."

Given that, could there be a more self-destructive, egocentric choice for a lifestyle? Mike, Skindog, Pete, and Evan are married with children. Gerlach has a steady girlfriend whom he loves dearly. Every time these guys paddle off into the unknown, you could argue that they're embarking on a quest no less daunting, deadly, and selfish than that of *Into the Wild*'s Chris McCandless or the crew of the *Andrea Gail*.

But how much of a *choice* is it? You might say, well, the crew of the *Andrea Gail* had to feed themselves and fishing was all they knew, so they had no choice but to motor off. And yet, if they really wanted to, each man could have found less-dangerous ways to make a living. In fact, I see real parallels between Parsons and his friends and Sebastian Junger's ill-fated crew. I'm fortunate to share a dingy little office at a fishing dock outside Charleston, South Carolina, and I have regular occasion to shoot the shit with a crew of tight-knit fishermen. When the crew of the *Hailey Marie* or *China Girl* come in after being slammed around in a nor'easter or battling a fourteen-foot tiger shark, they're exhausted, but they're also *lit up*—energized and alive. And they are no more constitutionally fit for an office job than Snips is for selling real estate. Sometimes they hate their jobs, but more often it's obvious that there's nothing they'd rather be doing. Such muddy, bloody work is *chosen* because they'd be lost without it, and without the ocean.

When I asked Parsons why he surfed big waves, he first went back to his childhood. To the heckling he endured at school, to his love of competition, to his almost insatiable need to be the best surfer in the water, and to a gnawing need *to stand on the shoulders of giants*. But that was as far as he could go. He didn't really know where the drive originated—only that he had it.

I asked Skindog and Pete Mel. Both said part of it had to do with the mutual pushing of the wild-assed circle of friends they grew up with—a group of surfing hellions known as the Santa Cruz Ratpack. But that didn't explain everything. "Sometimes I bust on, *why am I doing this?*" says Skindog. "Well, I enjoy it. Do you enjoy it to the point of dying? No, but I do enjoy it. So why do I keep doing it?"

Wherever the impulse comes from, Skindog and his lovely wife, Annoushka, are already seeing similar traits in their six-year-old daughter, Aianna, and tow-headed two-year-old son, Koa—especially Koa. On the day we spoke, standing in their kitchen, Koa was sucking a bucktooth-shaped pacifier and hurling pieces of his model train at me. "Quit it!" Skinny yelled, scooping him up and giving him a big tummy raspberry. "The kid scares me," he said in a conspiratorial tone. "He's already stuck his hand in a stove vent and nearly pulled his finger off. Next day, he's doing the same thing with his other hand. I watch him and say, 'What's wrong with that guy? Is he stupid, or programmed for it?'"

I think back to Brad Gerlach and his reckless, high-diving father and to the manic, driven competitive energy of Mike's father, Bob. I tell Skinny, maybe Koa's programmed for it, because maybe Dad is too.

In the late nineteenth century, Sigmund Freud was the first person to posit a genetic component to what we might call "thrill-seeking" behavior. He suggested that there were "torpid types," who felt best at a low level of excitement, while "vivacious types" craved excitement. But it wasn't until the 1960s that a curious young psychologist at the University of Delaware named Marvin Zuckerman took Freud's ideas and put them under the microscope. He asked test subjects a litany of questions: Do you like to drink, gamble, or experiment with drugs? Does monogamy make you happy—or would you prefer multiple sex partners? Might you enjoy leaping out of an airplane "just for the thrill of it"? Around 17 percent of humans, he found, are drawn to risky, sensory-stimulating behavior like moths to a flame. They also tend to be less religious, laugh more, dominate social group settings, and love testing exotic foods.

Zuckerman coined a term. These people were *high-sensation seekers* (sometimes abbreviated HSS). In short, they're prisoners, slaves, and addicts to adrenaline.

Zuckerman is quick to point out that sensation-seeking behavior runs along a complex continuum, and no two sensation seekers are alike. But when it comes to potentially deadly activities like big wave surfing, there are common threads. "Part of it goes with how you expect to feel in a situation," he told me in a telephone interview. "Even the risk of dying. If you expect to feel very anxious, it's not very pleasant to anticipate a sensation. But with the high-sensation seeker, the anticipated possible sensation outbalances the fear of dying." When I asked Zuckerman about big wave surfing, specifically, he said there were real similarities between guys like Mike Parsons, Travis Pastrana, and a hellbent skydiver. "Even before they do something for the first time [such as landing a double jump, parachuting, or dropping into a massive wave], high-sensation seekers also tend to estimate a risk as lower than low-sensation seekers. After they've done it a few times and survived, the risk appraisal is lowered even further still. They feel confident they can handle a situation."

A couple of decades back, Mark Foo famously said, "If you want to ride the ultimate wave, you have to be willing to pay the ultimate price." Many dismissed Foo's remark as glib, even cynical marketing hyperbole, but was it really hype? Really, Foo expressed a genuine truth that all his big wave surfing friends understood implicitly: A major component of the thrill that draws them is risking death. What few understand is where that truth and impulse comes from.

All the factors that lead to HSS behavior are not yet known, but one source is clearly, and genetically, biochemical. In 2006, researchers discovered a gene, called D4DR, that helps regulates the body's release of dopamine, a neurotransmitter that helps keep us on an even keel. Those with an inordinately long version of this so-called sensation-seeking gene possess an exquisite, almost bipolar sensitivity to blood dopamine levels. At what most of us would consider normal dopamine levels, HSS subjects feel low, even depressed. When a heavy situation triggers a dopamine release, glands squirt out adrenaline, endorphins, and estradiol. The blood becomes a nitroglycerine stew, and this lights up pleasure centers of the brain, yielding a high those who have never tried crack or heroin would find difficult to conceive. "Adrenaline is actually a peripheral," said Zuckerman. "It affects the muscles. The main neurotransmitters are noradrenalin or norepinephrine. When those are aroused, the whole brain is activated. Then dopamine also creates arousal—a pleasurable arousal. The heart pounds. Sugars are released. You can have the same level of heart rate and blood pressure increase as you would during an orgasm."

And just like a drug addict, over time the sensation seeker often requires a bigger and bigger rush for a fix.

Parsons, Gerlach, Skindog, Slater, and Mel find Zuckerman's ideas fascinating—particularly as they relate to fear and danger. For example, in most humans, a startling event—someone firing a gun nearby or jumping out and shouting "boo" from behind a door—would create a panicked increase in anxiety and heart rate. The high-sensation seeker will actually experience a *decrease* in heart rate and might actually find the experience *pleasurable*. At very high levels of danger, the sensation seeker becomes highly calculating, thinking several steps ahead, working out various survival scenarios at lightning speed. He becomes calm, even euphoric. Every sense is amplified. I've interviewed a few marines who are veterans of the Iraq War and who also surf big waves. They actually describe combat in a similar fashion. Of course, Jim Houtz described something very similar in himself while trying to motivate his fear-frozen crew on the deck of *Jalisco*.

"I remember everything in big waves," Skindog says. "Every *single* thing. I get totally clear. You get this wide sight—you can see *everything*. When you put your hand down in a wave, you feel *everything*."

Sensation seeking can be maddening, too. Some come to need either constant access to or the promise of impending danger just to get through the day. When he was younger, Evan Slater's wife, Jennifer, found him almost impossible to be around whenever there was a chance Maverick's might break. He insists that age and fatherhood have mellowed him a bit, but not entirely. "I've never been a fighter or very confrontational. But one thing I always did enjoy was going into a mosh pit and getting my ass kicked. Coming out of that at the end all bruised and beat up and, like, hugging everybody, going, *yeah*, that was *sick*. Big waves are kind of the same thing. Way more of a rush, and probably more healthy overall. Today, knowing that at any time I could go grab my nine-six [9-foot, six-inch surfboard] and put myself in that position is a good feeling. I have no problem being the guy to go and shop for gymnastic leotards for my daughters—all that domestic stuff. If that was all I did, I'd be pretty depressed. But big wave surfing makes me like, *Super Mister Domesticated*."

Skindog's wife, Annoushka, calls her husband a *complete* addict. "If he knows a swell is coming, he's just so amped the night before," she says. "He'll have so much energy. Like a kid waiting for a Christmas present. I'll be like, *simmer down*."

"My heart's beating a hundred miles an hour," Skindog says. "But as you get older, too, you get conditioned for the hangover—the adrenaline hangover—that comes from a weekend of riding big waves. I've seen it with my friends. Pete gets super grumpy after a big swell. No matter how high you get, the bigger the comedown. That's why you need the high again."

Gerlach was the offspring of a stuntman *and* a stuntwoman, so his future was pretty much foregone. A high-sensation seeker is simply what he is. It's part of every decision that's made through their whole life, for better or worse. "You know, I was one of the best guys on the World Tour," he says, eerily echoing his dad. "Sometimes I think, *I could have just stuck with it and been financially secure.* But I couldn't. I just *can't* be bored.

"Surfing these *huge* waves—I never thought it was something I would do. I've really had to rely on Mike telling me, 'You're good enough to do this.' I am? Really? I can't even believe I'm here right now. I'm picturing the enemy. I've had to conquer so much fear. I've had some of the most amazing days of my life conquering fear—seeking sensations."

"I think about my motivations," Parsons says. "God, it's funny sometimes. I went and did the highest bungie jump in the world in South America. I went with Taylor Knox, Kelly Slater, and one of Kelly's other friends. The funny thing was that me and Kelly were probably the most reserved—the most calculated. Taylor, he just went to the edge and jumped. I mean, I don't fit the profile of the guy who will go and skydive—I'm actually pretty conservative. I'm actually terrified of certain things. I drive slow, I'm always worried when I'm on the road. I fly airplanes now—and I'm super by the book. You'd think I'd be the guy barreling in the clouds, but I'm not. I triple-check the weather. I'm super calculated."

In a way, Parsons might claim Starbuck, the whaler in *Moby Dick*, as a cousin. As Melville wrote:

> "I will have no man in my boat," said Starbuck, "who is not afraid of a whale." By this, he seemed to mean, not only that the most reliable and useful courage was that which arises from the fair estimation of the encountered peril, but that an utterly fearless man is a far more dangerous comrade than a coward.
>
> "Aye, aye," said Stubb, the second mate, "Starbuck, there, is as careful a man as you'll find anywhere in this fishery. But we shall ere long see what that word 'careful' precisely means when used by a man like Stubb, or almost any other whale hunter."

Implicit in Parsons's world of calculated risk is the idea that somehow because he looks before he leaps off the highest bungie jump in the world, triple-checks the weather, and flies by the book, that he is less of a high-sensation seeker, or risk taker, than his friends. He's not. Flying by the book doesn't mitigate the fact that if he loses an engine above the hills of San Clemente, he's dead. And why does he fly, anyway? It's so he can surf big, empty waves along Baja's loneliest, most dangerous stretches of coastline—far, far from any medical attention. Parsons might be careful. He might be calculating. But like Starbuck, he deliberately chose brave yet terrified crewmen like Chris Mauro and eventually Brad Gerlach to go with him, and he still hunted the whale.

If Mike Parsons is an addicted big wave hunter, he and Starbuck share another trait in their earnest and general sobriety. Parsons always managed to turn his daredevil addiction into the hunt—something physical, an activity that provides a recognizable living, even if it's not as socially and commercially productive as, say, harpooning whales. For Peter Mel, on the other hand, the big wave hunt was not enough. This was something I learned of on February 14, 2008, while I was perched up on Rob Brown's boat watching Greg and Rusty Long, Skindog, Evan Slater, and Pete Mel drop bombs at Maverick's. Mel went ass over elbows on a huge wave. He paddled over to the boat. He was *done*. Rob Brown offered him a painkiller.

"No, thanks," Mel replied.

"How 'bout a beer?"

"Nope. Don't take anything anymore."

"How come?"

"Because I was a fucking drug addict."

Pete Mel comes from a line of watermen. His grandfather Peter was a fisherman. His dad, John, started shaping surfboards in the South Bay of Los Angeles in the 1960s, and he eventually moved to Santa Cruz, where he set up a shaping business and opened the Freeline surf shop—a place that's now been in business for forty years. When Pete was six, his dad pushed him into his first wave at Sewer Peak in Santa Cruz. Pete's board pearled, and he became entangled in slimy kelp. The experience so terrified him that he didn't try surfing again until he was thirteen. That time though, the hook set, hard. Pete, Skindog, and the Santa Cruz Ratpack would become a force to be reckoned with in the powerful waves around Steamer Lane. Pete was big, close to Laird Hamilton in stature, and his rugged frame served him well in bone-crushing surf. His graceful, swooping turns inspired his nickname, "The Condor." Eventually Pete and buddies made pilgrimages to the North Shore of Oahu and were introduced to Maverick's in the early 1990s.

Pete met his future wife, Tara, in 1991 at a party. She was tall, drop-dead gorgeous, smart, and funny to boot. Pete was a hilarious guy with a grand personality. "He just lit up a room," she says. "But he was also very kind. My son was one year old. Pete was so sweet to him. Played with him in the water. Taught him how to surf. My ex was never really in the picture, and I remember one day my son called Pete 'Daddy.' I said, 'Is that okay?' He said, 'Actually, I like it.' I was like, 'Okay, he's a keeper.' I felt safe and very protected around him. I still do."

At the turn of the millennium, Pete Mel was at the top of his game—Evan Slater called him the best all-around big wave surfer on Earth. He appeared alongside Skindog, Darryl "Flea" Virostko, Brad Gerlach, and Mike Parsons, all naked, in a *Vanity Fair* spread. Yet he was, he says, "a selfish motherfucker."

For Mel, the substance abuse began with drinking during the big wave off-season. Something to keep up the buzz during the long, frustrating months when the surf was down. "It's funny," he told me later. "The best part of it all is actually when a big wave session's all said and done. Sometimes the buzz lasts a day. Sometimes a few days. It's euphoric. Just something you can sit well with. Like, *yeaaahhh*. But then eventually you get a comedown and the soreness will kick in. Sometimes I'll get sick. You wanna keep it going. Maybe a drink and a cocktail to keep up the buzz. Then other drugs, too, unfortunately."

One night when he was fairly well lit up, he snorted a line of cocaine. It lit him up further, so he tried a little more. "I've always overindulged and done stupid things," he says. "It runs in my family."

Mel thought he was keeping the powder and sauce a secret from Skindog and Tara, but he gained forty extra pounds, which is tough to hide. "I was watching Pete and a few guys and going, 'Are you *fricking kidding me*?'" says Skindog. "Especially Pete. He has a wife and kids. For me, my family is my cocoon. I always enjoyed having a good time, but once you're not having a good time, it's done, and I don't let anything like that *near* my house. So I'm like, 'Pete, dude, what's up? Oh really? You're doing a gram of coke a day and drinking forty beers?'"

It was simple. Big waves made Mel high. When he didn't have that high, he craved it. Drugs and booze became a crutch. Mel went cold turkey and stayed clean for a while, but he remained frustratingly overweight and out of shape. Maybe, he thought, a little speed would get him back on track. He snorted some methamphetamine. Now *this* was something. It took away his desire to drink. He started losing weight, surfing a lot, getting shit *done*. He was soon hitting it all the time—*all the time*. Before long, he could once again survive a two-wave hold down at Maverick's and come up laughing.

"You could see it in Pete's head—he was going a million miles an hour," says Skindog. "*Spun out*. He started getting all skinny. He's like, 'Yeah, I cut back on the red wine.' Then he disappears. I can't get hold of him. When I did see him, I could just see him looking at me ashamed."

Skindog's buddy and three-time Maverick's surf contest champ Flea Virostko was also getting in very, very deep.

One day, Mel began to wonder if the house had shifted on its foundation. When voices began speaking directly to Mel through the box on his cable TV set, he cut the cable to quiet them. Shortly thereafter, he sat at home with his head in his hands after a major bender with Flea. Tara put her hand on Mel's shoulder. "I said, 'Honey, are you okay? Have you been using some drugs that could make you paranoid?'"

Mel looked up at Tara through bleary eyes and said, "Yes."

"It makes me kind of sick to remember," she says, dabbing her eyes.

That moment became Mel's wake-up call. Forty years ago, his grandfather heeded a similar call and has remained sober to this day. Back in 2002, though, meth killed Mel's aunt—a fact he wouldn't learn until after he'd made his own decision to quit. "I've had other family members who didn't get it," he said. "It took me a while, but I got it."

Mel's been clean now for nearly six years. He's opted to channel his sensation-seeking genes back into big wave surfing—and keeping himself in top shape. As of this writing, Flea's been clean for better than two years. He's put his focus back into big waves, and into a rehab program for guys like him. He calls it "Fleahab."

Tara's been left in the terribly difficult position of having to hear the awful things Flea and her husband did, as they are rehashed publicly time and again. Both surfers told their stories for an article in 2009 in *Surfer*, titled "Coming Clean" by Kimball Taylor. Taking the fifth in the twelve steps—admitting "the exact nature of our wrongs"—Mel and Virostko voluntarily recounted everything from Flea's literal fall from a cliff that should have killed him to the times he and Mel served both their addictions to drugs and big waves *simultaneously*. They revealed the grisly depths of their addiction to a drug—meth—that Taylor called Santa Cruz's "gorilla in the room" in the hopes that, maybe, they could keep others from sliding down that same icy slope.

The story garnered a great many online reader comments. "Some people had some nasty things to say," Tara says. "But most were really grateful and thanked Flea and Pete for the article. Then there was one guy—it was just one sentence."

She pauses. Her eyes again well with tears and her voice breaks.

"It said, 'I think you just saved my life.'"

Chapter 9:

ON THE
SHOULDERS
OF GIANTS

"Here comes one scurvy type leading another!
God pairs them off together, every time."

—Melanthios, from Homer's *The Odyssey*

By January 2001, Larry "Flame" Moore had spent just over a decade being thwarted in his attempts to document a surf session again out at the Cortes Bank. In that time, he and Mike Parsons had launched at least three weather-aborted paddle surfing missions, and he and Mike Castillo had flown out over Bishop Rock perhaps fourteen times. They buzzed the waves at rooftop level on west swells and north swells, during long periods and short periods, and relayed their observations back to Sean Collins. Flame returned puzzled every time. Though they found big waves, they were never the eight-story titans they had witnessed during the seminal Eddie Aikau swell of 1990.

Then on January 14, Flame rang up Parsons. A 956-millibar storm was plodding across the Pacific. If the forecast held, hurricane force winds would soon be raking a thousand-mile swath of ocean between Hawaii and California. "I'm putting out the yellow light for this Cortes thing," Flame said.

On January 17, the forerunner waves swept beneath NOAA's Southeast Papa buoy, a storm-tossed distant early-warning system six hundred nautical miles west of Eureka, California. A solid twenty-second swell was two days out. According to Collins's calculations, the angle of approach looked ide al for Cortes to work its magic.

Still, on January 18, Collins began sweating bullets. The winds on the Tanner Banks buoy had been bad out of the northwest all week long. Cortes was perched even farther out, so conditions would be even windier and choppier. The mission called for an expensive armada of sea and aircraft and Flame told Collins that if they blew it, they wouldn't have the budget for the trip again that year. "There was very serious thought to pull the plug on everything and saving the expense of doing the trip in bad conditions," says Collins. "Needless to say, it was a very nervous night of sleep. I could hear the swell build overnight on the beach with the long-period 'thumps.'"

Yet no one would know for sure until they were out there—bobbing in and hovering over MacRae's Rock—if the wind would lay down enough so they could harpoon Flame's elusive white whale.

During the last decade, big wave forecasting had matured a great deal, beginning with Collins's first reliable forecasts for Todos Santos, but it was still exceedingly complex. To begin with, NOAA's data buoys weren't really telling the whole story. Your weather radio report might state that a swell hitting the Half Moon Bay buoy was 20 feet at twenty seconds—an earthshaker. Yet that swell might actually be a combination of a 5-foot, twenty-second westerly groundswell and a 15-foot, six-second southerly windswell—a comparatively miniscule event. Data buoys actually transmit data on *all* swells hitting them—a remarkable bit of scientific wizardry—but NOAA doesn't make that data available to the public. If Collins and Flame wanted consistent, pinpoint forecasts for the Cortes Bank—or any big wave spot—they needed to sift through every swell in the water.

To help do this, Collins hit up a buddy who developed computer code for NASA named Jon Chrostowski. They eventually created a program that hacked those buoy data streams. "It was such a bitch to figure out that code," says Collins.

The second piece of the Cortes puzzle lay concealed along the tortured, craggy seafloor off Bishop Rock. Collins partnered with a Scripps Institute Oceanographer named Bill O'Reilly, whose doctoral thesis involved modeling incoming swells around different bottom topographies. The Bishop Rock fascinated O'Reilly. At around six hundred feet, he and Collins noted a peninsular thumb of sandstone and basalt that juts out three to four nautical miles to the northwest of Bishop Rock. A long-period wave, something in the eighteen- to twenty-second range, carries energy down beyond a thousand feet, but the energy really begins to concentrate at around six hundred feet. If a wave swept toward the Bank from the proper northwesterly angle—somewhere between 270 (true west) and 360 (true north) degrees on a circle (though it should be noted that Collins keeps the best combinations of period and swell angle a secret)—the wave *should* hook onto that thumb, slowing and bending inward on itself and curving toward the shallows. The Cortes Bank's ancient terraces would then cause the wave to shoal steadily, while the hook would focus even more energy onto the Bishop Rock like sunlight through a magnifying glass. The wave would rise higher and higher in the shallows until reaching the final big stair step. There it would trip up, careening and falling forward like an enraged giant while peeling down the shallow waterline like a line of toppling dominoes.

Collins and O'Reilly marveled. If they were interpreting the bathymetry properly, a 15-foot, twenty-second swell might grow to four, maybe five times its height as a breaking wave. With a pure swell of 20 to 25 feet, the comically impossible would become real—a peeling wave 100 feet high. Fifty feet of pure groundswell—perhaps a once-in-a-century event at these southerly latitudes—might create a 200-foot wave. If you could generate *enough* swell, a reeling, Malibu-style wave 1,000 feet high was theoretically possible.

To date, this is a combination of factors that exists at no other big wave spot yet revealed. Waimea, Maverick's, Todos Santos, Outer Log Cabins, Jaws—all seem to have upper limits between 50 and 100 feet. Above that, a swell would simply rear up along the outermost reefline into a terrifying and unrideable wall that would close out along its entire length. In fact, Maverick's is known to have done this on at least a couple of occasions.

"Cortes is a ridgeline," says O'Reilly. "The crest has an orientation that lines up off to the northwest. Where a wave breaks depends on the size of the swell. You have this natural slope that can take just about anything thrown at it."

Really, anything? I ask. What about an earthquake-generated tsunami? O'Reilly and Dr. Gary Green, a professor at the Moss Landing Marine Laboratory who has worked on exquisitely detailed sonar-mapping projects of the Bank, point out that a tsunami is fundamentally different. A huge, storm-driven Cortes wave might be 100 feet high, while the tsunami that hit Japan in March of 2011 was, by most accounts, 60 feet at its highest. Yet rather than, say, a sixty-knot, mile-wide storm wave with a twenty-second period from trough to crest, a tsunami might be a four hundred-knot wave *one hundred miles wide* with a twenty-*minute* period, and thus countless orders of magnitude more power. The effect at Cortes would depend on how far away the tsunami was generated and the force that spawned it. There might only be a violent surging and swirling of waters and a slow rise and fall of only a few feet of sea level, or the results might rival a science-fiction movie. "If you have 60-, 80-, 100-foot waves out there just from a storm, and you have a big tsunami in the Aleutians that comes up on that island, it's going to build up at least that high," Dr. Green says. The Cascadia subduction zone—a massive fault stretching from southern Canada to northern California—tends to generate apocalyptic tsunamis every three hundred years or so. Over millennia, particularly when Cortes Bank was Cortes Island, plenty of such science fiction waves likely swept entirely over it.

Naturally, the enormous January 2001 swell that was brewing caught the attention of big wave surfers everywhere. In Northern California, Jeff Clark served notice on January 17 that the Maverick's Invitational contest might run in two days. This presented a problem. Pete Mel, Skindog, and Evan Slater were expected to show in Half Moon Bay for the contest, but Flame wanted them at Cortes.

Further, Mel and Skindog had promised filmmaker Dana Brown they'd surf the Maverick's contest for his film *Step into Liquid*. On the morning of the eighteenth, Mel's phone rang. "Hey, are you going to Maverick's tomorrow?" Brown asked.

"Uhh, I'm kind of doing something else, and I'm kind of sworn to secrecy," Mel said. "Let me call you back."

Mel nervously called Flame. Was it okay if Dana came out to Cortes? "I don't care what he does," Flame said.

Mel then called Jeff Clark. "Ummm, I have this other deal, to go down south," he cryptically told his old friend. "Mavs is supposed to be windy and, umm, not that good." Mel laughs about it today. "I don't remember actually trying to bait Jeff into not running the contest, but I told him, 'if you're *not* running the contest, *I wanna do this*.'"

Clark decided to delay his Maverick's contest, and Mel called back Dana Brown. "We're going to surf the Cortes Bank, a hundred miles out to sea. Get yourselves a boat."

Once Flame made the decision to go, everyone flew into action, preparing for the trip. The first order of business for Flame was ensuring their boat and its crew were ready. He and Evan Slater had earlier lined up Gary Clisby, a former pro surfer and professional sportfisherman who owned a fifty-one-foot charter vessel called *Pacific Quest*. When the word was given, Clisby assigned his twenty-one-year-old captain John Walla to lead the effort. Walla didn't have much notoriety, but Parsons and Slater knew the rugged youngster was one of the most driven, hard-core young watermen in all of California. As a surfer, he charged Todos Santos just as hard as they did. Maybe harder.

Walla was over the moon. He called his buddy James Thompson, a young man he regularly terrified with hair-ball diving, climbing, spearfishing, and surfing adventures. Walla offered him a first mate's position for the trip. "We're going to the Cortes Bank to surf," Walla told Thompson. "You in?"

"What the hell are you talking about?" Thompson asked. His new boss at a day trading stock office was standing over his shoulder.

Walla said, "Bring your big board. We'll paddle in."

Thompson turned and apologized to his boss. Even if it cost him his job, he *had* to take a vacation day.

Walla had captained plenty of tuna fishing excursions along the inner waters of the California Bight, and he was essentially, frightfully unflappable. He was also familiar with the open-water conditions around the Cortes Bank, a place he only half-jokingly referred to as "The Flemish Cap."

"It's just a different ocean once you get outside of Point Conception," says Walla. Waves of albacore appear between San Clemente Island and Cortes in June and July, "but that's when you *don't* want to go out there. The wind's howling, it's freezing. It's a lot more like being in Northern California. It's the *real* ocean. You see whales everywhere and these huge baitballs of anchovies. Albacore and finback whales—they come a foot from the boat, just tearing through them."

Walla set about ensuring the *Pacific Quest* was outfitted. Clisby told him they'd be hauling a trio of thousand-pound WaveRunners, and they would have to use the boat's sketchy hoist to lift them aboard. That, Walla thought, would be interesting.

Meanwhile, those who would not be going were also getting the news. At the *Surfer* offices, rumor of the mission reached Sam George late on January 18 via lensman Jason Murray. "Jason said, 'Look, I can rent a plane and fly out there and shoot it,'" George recalls. "He got all excited. Then Larry, my beloved Flame, got word of it and called me up."

"I spent ten years on this Sam," Flame ranted. "That's *my* wave. I can't believe you wanna poach this mission."

"That's the word he used," laughs George. "*Poach.*"

George let Flame spew, then finally interjected, "Larry, can I speak for a second? If you're asking me, as a magazine editor, not to cover this story, I'm going to say 'fuck you.' But if you're asking me, as a friend of twenty years who has shared many adventures with you, then I will do so.'"

The line was silent for a moment. "Sam, I'm asking you as a friend," Flame said.

George replied, "Then Larry, we won't go to Cortes."

George kept his word and didn't go. Neither did George Hulse, who despite being on the original 1990 Cortes trip had moved away from pro surfing and helped found the now popular Shoreline Church of San Clemente. Yet, at the very last minute, the one person Flame didn't forget was Bill Sharp, the man who, as the author of "Project Neptune," had dreamed up this mission in the first place. No one had wanted Sharp's *Surf News* to break the story, and Bill held no official ranks at *Surfing* or *Swell*. If he'd asked, Bill could have probably journeyed aboard the *Pacific Quest*, but the boat was relatively small, and it would be mighty crowded on board. He didn't have the dosh for his own boat, photographer Rob Brown had already motored off, and Mike Castillo had flown off to Baja to chase swells. Flame was going to go in a tiny Baja bush plane, which would be flown by another surfer named Vince Natali. Bill could have the spare seat.

The surfers Flame and Evan Slater invited were Mike Parsons, Brad Gerlach, Skindog, and Peter Mel. Per the usual routine, these men quickly let their girlfriends, wives, and in some cases children know that as of, let's see, tomorrow they were dropping everything to go surf. Whether their surfing life was considered a selfish choice or an addictive compulsion, it was never easy for their loved ones to live with.

Mel and Slater were both husbands and fathers of babies in their first year of life. Mels' wife, Tara, wasn't overly concerned. Mel had always survived these missions, and it was easier for Tara if Mel simply didn't reveal too much.

Skindog's wife, Annoushka, on the other hand, was freaking out. It didn't help that her brash husband was typically nonchalant.

"You're just pissed that I'm going on a surf trip," Skindog told her.

"No," Annoushka replied. "That's not it. I'm pissed because you're going on a surf trip to try to surf 100-foot waves—with a pretty good chance of dying. Nobody's ever done this. It's uncharted territory."

"I'll get life insurance," Skindog replied.

"That is *not* what I want to hear," she shouted.

On the morning of January 20, the day they were going to leave, Mike Parsons was still frantically trying to get gear in order. Parsons was the planner, and a tow-surfing mission a hundred miles out was not like a jog down the path at Trestles.

Parsons latched his cell phone to his ear and began a panicked rundown with Brad Gerlach. They would need spare rope, tie-downs, bungie cords, carabiners to lash the rescue sled to the WaveRunner, surfboard leashes, five cans of gas, two-cycle oil, spark plugs, jumper cables, wetsuits of varying thicknesses, neoprene booties, gloves, hoods, life jackets, walkie-talkies, a spare ski tow rope, anchor, extra foot bindings and screws, lead weights, an assortment of fins, several bars of surfboard wax, and, oh yeah, a 10-foot paddle surfboard—just in case.

"I couldn't believe how frazzled he was," Gerlach told Evan Slater. "When we were growing up doing contests, it used to piss me off how prepared he was. He had backups for his backups. Now when I'm thinking I can actually benefit from that, he's like, 'Hey, do you have a leash I can borrow?'"

Mel and Skindog reached San Diego Harbor at 11 A.M. They were anxious to get their WaveRunner in the water. But the long, bouncing drive had dislodged the ski's exhaust manifold. When Skindog pulled away and motored toward *Pacific Quest*, the ski began to sink. Skindog screamed for Mel. By the time Mel made it back to the ramp, his buddy was stripped down to his boxers and struggling to keep the drowning machine's head above water. Then a kid rolled to the ramp on a skateboard.

"Hi," the kid said. "I'm Johnny. I'm going to take you guys out to the Cortes Bank."

Mel raised an eyebrow. "So you're the—*deckhand*?"

"No, I'm the captain."

Skindog laughs. "It scared the hell out of me.'"

Walla was scared, too, but for a different reason. Watching them wrestling their sunken ski out of the water, he thought, *Oh my God, these guys are gonna die out there.*

Skinny and Mel rescued the ski and were soon met by Slater, Gerlach, Parsons, photographer Aaron Chang, his ski driver Randy Laine, Chang's backup photographer Brendan Hayes, and videographer Fran Battaglia. Bro handshakes were exchanged all around. Flame remained in the *Surfing* offices, with a phone plastered to his ear frantically directing last-minute details, while also keeping an eye on the light box—doing his mundane day-to-day duty of picking photos for the next issue.

Dana Brown arrived with his film crew; they would venture out aboard *Pizzazz*, a sportfishing boat he had hastily managed to line up. The most impressive item in Brown's arsenal was an enormous, gyroscopically balanced camera rig that would allow for a steady shot even in a heavy swell. Dana had to assemble his crew with such haste that he didn't even know everyone. He joked about the mission into the unknown with his newly minted assistant cameraman. "Then the captain started talking about how he's heard about Cortes and how scary and gnarly it is," Brown says. "But you know, he's a real good captain and not to worry."

The captain of Brown's boat then gave his safety debriefing on life jackets, life rafts, and fire extinguishers. As he spoke, Brown noticed that his assistant cameraman was slowly, steadily stepping backward out of the pilot house. Suddenly, he leapt off the boat. "Sorry to hang you guys up, but I can't do this," he said. "I have a *really* bad feeling."

He broke into a full sprint up the dock. "How are you gonna get home?" Brown yelled.

"Don't worry about it," came the reply.

Brown turned to his production assistant, a girl named Julian. "Well, that kind of freaked me out," he said.

Julian laughed nervously. "You think?"

The two boats departed San Diego at sunset. When the crew of *Pacific Quest* learned that bananas had been brought on board, they followed an ancient maritime good luck protocol and chucked them over the side.

Early that same day, photographer Robert Brown had driven his catamaran out to Catalina Island with a sailing buddy named John Connors, who also happened to be his insurance agent. The pair anchored off Avalon and paddled Brown's tiny inflatable to town. They had dinner and rented rooms in a local hotel, intending to land a few hours' sleep before heading out to join everyone at the Bank. Then around midnight, Brown flipped on his weather radio. The Tanner Banks buoy had suddenly lit up with something Jim Houtz would have recognized immediately. The swell height was a relatively small 8 feet, but the period was an astonishing *twenty-five seconds. It's going to be fricking huge,* Brown thought, *and there is no wind.*

Brown leapt out of bed. "Let's go," he told Connors.

They hoisted the inflatable above their heads, walked past the noisy bars of Avalon, and stroked back out to the catamaran. Brown hailed the Harbor Patrol. "We're going to the Cortes Bank," he said.

"You are?" Came the reply. "Well, good luck with that."

Brown followed his newfangled GPS around the backside of San Clemente Island. The military was conducting bombing and artillery exercises and the vast bombing range was lit up by huge flashes and laser-like tracer fire and wracked by deafening explosions. It looked like an attack from *Star Wars.*

The captains of all three boats had agreed to make contact on the VHF to keep one another apprised of their progress. Brown hailed them all through the lonely night as he motored toward the unknown. No one answered.

The surfers aboard *Pacific Quest* turned in at 10 P.M. They'd all had a beer or two but were still jittery. Parsons lay in a bunk in the bow, feeling the long, low swell build as the boat nosed farther into the open ocean. The Santa Ana wind was disconcertingly strong when they had left, but now it seemed to be laying down. He kept asking himself, *What the hell is it going to be like out there?*

At around 1 A.M., the sound of silence woke Peter Mel. He found Walla, Thompson, and deckhand Mike Towle assessing a situation. One of the fuel filters had apparently clogged, starving an engine. Walla replaced the filter, but the engine then refused to restart. An hour of checkouts yielded nothing. Perhaps there was air in the fuel line. Maybe a pump had failed. No one knew.

Walla fired the remaining engine and continued on. "He didn't yell at anyone," says Mel. "He never blamed anybody. And in the end, there was no debate. *We're going.*"

"Johnny, wake up. Wake up, dude."

Walla took a groggy look around the darkened cabin of the pilot house. It was 4 A.M. James Thompson's face was bathed in a soft, green glow. He was staring transfixed at *Pacific Quest's* tiny lithium radar screen. Eerie pixelated lines were flickering into existence, scrolling down, and then fading back into black.

Thompson was well spooked. A huge freighter had steamed across the black horizon, but her lights had been suddenly and then regularly eclipsed by some barrier between the boat and Bishop Rock. Then these ghosts began to haunt the radar screen.

"What the fuck is that?"

"It's the Bank," Walla replied, his finger following a line as it began a matrix-like descent. "Those are waves."

Everyone soon gathered round the radar screen, watching real waves detonate virtually in the invisible distance. Evan Slater saw a strange, symmetrical beauty in the imagery. Dana Brown's and Rob Brown's boats were tiny blips. The waves were anything but.

The sky still lay under a smothering blanket of stars and the ocean was an inky, infinite black. Walla crept northward, keeping the Bishop Rock to his northeast flank. They couldn't see a damn thing. Above the idle of the *Pacific Quest's* single engine came a crack of distant thunder followed by a long roar. The glass in the windows shook a little.

The radio crackled. Rob Brown crawled into position, too, keenly aware that other shoal spots on the Bank might spawn deadly rogues. "Where have you guys been?" he asked.

Ten minutes later, dawn's early light revealed a shimmering plume of spray. A Himalayan peak rose to life far off the bow. It was shaped like a great, volcanic cone—43 million pounds of water, terrible and unrideable. Its foam exploded an unknowable number of feet into the air and churned the surrounding waters into a 360-degree maelstrom of confusion. Then as the set swung closer to the boat, a second peak was revealed. A righthand wave stood majestically before throwing out a yawning barrel whose size was, again, impossible to estimate.

Pacific Quest took up a position near the Bishop Rock buoy, its bell clanging a lazy wake-up call. On the shoulder of the waves, the surfers saw that the real locals had already beat them out—a floating posse of argumentative sea lions with fins pointed to the sky.

Unloading the WaveRunners was a precarious affair that Evan Slater likened to moving pyramid stones. It had been tough enough to maneuver the hoist at the dock. The skis were raised via a pair of slings, and they were now swinging wildly in the swells. When Randy Laine's thousand-pounder was jacked into the air, its rear end slipped and crashed onto the deck of the *Pacific Quest*. Had the nose also slipped, the ski's next stop would have been the engine room.

The moment Skinny and Mel's ski splashed into the water, an amped-up Skindog sped off toward the waves without so much as a good-bye. "Fuckin' selfish Skinny," Mel laughs. "*I* wanted to be the first one out there."

Skindog edged in. Away from the boat and bell buoy, the air became still and quiet—seeming to suck up even the sound of the idling ski. The ocean's surface was lumpy and a bit confused, a condition surfers call "morning sickness," and there was a good deal of kelp. Suck a few fronds into the ski's jet impeller, and you might become a sitting duck.

A set began to shoal.

"It's 8 feet," Skinny said to himself. "No, it's 10. No. Ohhhhh *Jeeesus*."

He tore back to *Pacific Quest*. The waves were easily 20 feet from top to bottom and building.

Rob Brown was quietly freaking out. This swell was still coming on. The strongest, deepest long-period bands were still an unknown number of miles out. There was no way to know when they would arrive or how they would react with the bottom. At Maverick's and Todos, the seafloor produces a very pronounced takeoff zone that makes photographing from a boat a reasonably easy endeavor. If you know what you're doing, you can safely hang a hair's breadth from the edge of disaster. This wave, though, was dangerously shifty—like Sunset on Oahu. Different sets seemed to focus on different sections of the reef, and Brown had seen waves capping on other distant shoals—Maverick's-size peaks. As energy filled in and tide dropped, waves might shoal across other spots on Bishop Rock where nothing was showing. No one really knew what lay on the bottom, either. Brown knew that the chart guide map showed a wreck to the inside covered by a mere three feet of water, but he had no idea where it really was or what that really meant. Nobody did.

One of the few who might have told them wasn't there. Santa Barbara diver Ben Wolfe has dived the *Jalisco*'s wreck probably six times. Once, he was even left stranded for a day on the Bishop Rock buoy when the current swept him away from a dive boat whose crew didn't realize that he'd gone missing.

Wolfe and his friends still dive inside *Jalisco* in search of lobster, reaching as far into her decaying innards as an engine room that leaks oil to this day. She's now in at least three pieces, and Wolfe is reasonably sure that a big portion of the concrete behemoth has been pushed just inside the surf zone. "There's fifteen, twenty feet of water over her collapsing deck, and she's not going anywhere," he says. "She's covered in marine growth, and barely recognizable as a ship."

Worse, her concrete is breaking down, battered by waves and eaten from the inside as rust consumes her rebar like a plague of metal termites. As more concrete crumbles, more jagged rebar is exposed.

"It's pretty crumpled down now," he says. "And there are just a couple of portions where you can get in. When you do get in, it's still pretty enclosed. But there are more and more holes in it—some blow through all the way inside now. You're diving in rebar all over, and you gotta pay attention to where you're going so you don't get lost. When you're inside it's calm, but coming in and going out can be rough. It's scary. I can't believe guys surf out there."

As far as anyone—even John Walla and Evan Slater—knew, only the tow-surfing teams would actually be riding waves: Skindog and Peter Mel on one ski, and Brad Gerlach and Mike Parsons on another. As the watercraft pulled away from the *Pacific Quest*, deckhand Mike Towle's words resonated. "Remember, everything's big about the Bank," he said. "And if someone needs help, keep in mind that it's a forty-five-minute chopper ride out—and the Coast Guard won't even bother if you're already dead." As if on cue, Flame, Sharp, and Natali appeared on the horizon in their plane. The weather was perfect, the view stupendous. At sea level, waves can appear somewhat flat and two-dimensional. From a thousand feet up, you could see the swells slowing and bending around the Bishop Rock, their three-dimensional spines literally tracing the contours of the ancient shoreline. Endlessly complex physics caused each to shift and warp in its own unique, almost hypnotic pattern. A mile to the north, Evan Slater's "Himalayan" wave indeed looked like a giant volcano. It was, said Sharp, "utterly fascinating."

Randy Laine had motored to a position inside of and directly in front of the breaking waves with his massive four-passenger Yamaha. Directly behind him sat Aaron Chang cradling a Canon with a big 300mm auto-focus lens. Laine was the first person to ever tow a human on a surfboard behind a Jet Ski. He had skied nearly every big wave on Earth, from Todos to Maverick's to Jaws and Outside Log Cabins—and he was spooked.

Gerlach and Mel were to grab their respective ropes and surf first. There seemed two spots along the reef where you could slingshot a surfer onto a wave. The teams approached gingerly, opting to get a feel for the seemingly safer inside sections before heading farther out. From a distance, everyone had marveled at how the waves seemed to topple in slow motion. Up close, the impression was completely reversed. Mel thought the wave moved two notches faster than anything he had ever seen.

Today, no one seems to remember who rode the first wave, but the consensus is that it was Peter Mel. Skindog picked a fairly small wave—what he later called 15 to 18 feet, in surfer terms, but which anyone else would call 20 to 25 feet, trough to crest. He chased it down from behind and expertly piloted Pete Mel a good position.

The glider pilot let go of his tow plane. Mel made a few short S-turns to get a feel for his 7-foot 2-inch JC Hawaii. It was a little stiff, but felt *gooood*. The wave let Mel just kiss its lip, then Mel arced a hard angle to his left and plunged downward. At the bottom, he laid every ounce of his two hundred pounds into a solid bottom turn and rocketed back up the face. He carved the wave like a Super G skier.

Next, Mike Parsons slung Gerlach onto an A-frame-shaped peak. Gerlach faded left, turning his board back toward the descending white water to tap into the precise spot where he might harness the wave's maximum energy, and then at the last instant, he laid a gauging right-hand slash across the wave's face. Gerlach was a technician; he liked nothing better than to lay his board down on a rail and draw out a turn as long and hard as he possibly could. Yet he had never in his life felt anything like the power of these waves—had never drawn out a turn so long and hard. It was like snowboarding down the steepest, deepest powder run on the planet while the mountain tried to eat you.

For the moment, the surfers maintained their roles. Parsons and Skindog towed Gerlach and Mel for maybe a half an hour. Bravery increased, and before long, the boys were edging their way farther up the point toward the northernmost peak. Directly in front of the waves, Laine and Chang were scared out of their wits. Laine kept having to jump into the water to pull clogging kelp out of the impellor. Rob Brown inched his throttle forward and followed warily in his catamaran, with Dana Brown's sportfisher tucked a little farther to the inside. Rob's buddy John Connors was hyperventilating. "I was spending 90 percent of my morning defending the fact that we were safe," Rob says. "He was so freaked out that I made him go sit down in the front of the boat. He was stressing me out so bad that I had a big knot in my chest."

Of course, Rob should have listened to his friend.

Parsons hurled Gerlach onto the first wave on the northernmost peak. Gerlach charged down the line, and as it reached the inside, the wave began to clamshell above him. He was enveloped in a massive, yawning barrel. The biggest of his life. When the wave cleared its throat, a cannon blast of compressed air erupted him from its windpipe and he shot out, still standing, into daylight.

"I remember Brock [Little] telling me that he fell inside a tube at Jaws because Brian Keaulana told him to jump if he felt anything hit him," Gerlach later told Slater. "But he was pissed 'cause he said he probably would have made it if he hadn't jumped. I kept thinking about that as I pumped through these blind sections."

"Oh my God! Gerr!" Mel shouted to Skindog, who was driving their ski. "Dude, get me *into* one of those."

Rob watched Gerlach's wave in awe. Between shots, he had been trying to figure out—as best he could—from which point he might most safely shoot. Going around to the far side of the wave and filming from the east, off the top of the reef, was out of the question. Every so often, Evan Slater's wedge wave thundered clear into the edge of the surf zone. It was easy to see why the nineteenth-century captain of the *Cortes* once thought he was above a caldera.

A wave unspooled off the top of the point, wrapping onto the reef like a whistle string spun onto a lifeguard's finger. Rob watched mesmerized as it steamed across the shallows. He sensed his boat turning a bit and looked over his left shoulder. A deep blue wall loomed above the catamaran. The wave had swung wide. It was draining water down off the reef and sucking Rob toward its maw like a black hole. He screamed, simultaneously yanking the wheel to the left and slamming the throttle forward. The boat nosed into the wave's face and slid backward a bit, the propellers straining to gain a footing, a terrified Connors hanging on for dear life. The wave was feathering, and Rob had a moment to think, *I'm looking straight up at the ocean.*

The boat climbed to vertical and sailed into the air, its entire twenty-nine feet fully clear of the water. It could have come back down either way, but miraculously, it flopped forward on the wave's backside, obliterated from view in an explosion of spray. Later, when Rob watched the last moments of the *Andrea Gail* in the film *The Perfect Storm*, he was struck with a powerful sense of déjà vu. A half second slower on the draw, and the catamaran would have been pitched backward.

A second, bigger wave loomed through the clearing mist. Brown gunned the engines and pulled up alongside *Pacific Quest* in deep water. Flame's voice crackled across the walkie-talkie. He and Bill Sharp had seen the whole thing from Natali's plane. "That's the spot!" Flame yelled.

"I almost died," Rob said, clutching his chest.

"That's the spot!" Flame was nearly frantic. "*That's* the wave I saw before. Tell the boys to get over there! That's where they need to be."

Riding together, Randy Laine and Aaron Chang pulled up alongside Rob. "I've never seen anything like that," Laine said. "That wave was double, triple the length of your boat. You went straight up it."

"His big boat just looked like a tiny little Jet Ski going over a wave at Jaws—a postage stamp," Laine told me.

Rob Brown's gaze returned to the waves. Dana Brown's far larger sportfisher had been allowed to drift into almost the same spot. Rob watched incredulous as another wave lumbered onto the reef. He grabbed his VHF microphone and yelled to the captain to move his ass. The boat just sat there. Then he heard the captain say, "Oh my . . . Oh my God!" A cloud of diesel smoke rose as he finally throttled his engines, the boat nearly broadside to the wave. It disappeared completely in the trough and then wallowed, sideways, up its five-story face.

"We're going, 'He's dead, he's dead,' Rob says. "We're gonna have to go save these people. Then he *just* rolled over the top, going, 'Uuugggggghh!!'"

Rob shouted into the VHF. "What the hell were you thinking?"

"Oh my God," said the captain. "I couldn't get it in gear quick enough."

This was indeed Sunset Beach—juiced on a tankerload of steroids. At Sunset, most waves break off the northernmost top of the point and peel down from there. But some waves instead load up their energy onto Sunset's West Bowl. There, they grow, warble, and barrel in a most spectacular and unpredictable fashion. Surfers spend lifetimes trying to get Sunset wired. This crew had a few hours at best. As the tide slackened, the best waves were now jacking up onto the Cortes Bank's version of the West Bowl, a spot that would come to be called "Larry's Bowl." And they were still growing.

Aboard the *Pacific Quest*, Evan Slater and Captain John Walla marveled at waves that were just *achingly* beautiful. Eventually, Slater could bear no more. "If this was five years ago, before everyone had safety blankets," he told Walla. "We wouldn't think twice, we'd be paddling in, no problem."

"Okay, let's go have a closer look," Walla said. He knew full well that if they paddled out, they were going to get close enough to try and catch one. It looked intense out there, but it looked surfable.

James Thompson sat on the roof of the *Pacific Quest* casting a fishing line. Mike Towle was behind the wheel. Walla called up to Thompson. "Hey, we're gonna put on our suits and paddle out. You in?"

"Umm, no," Thompson said.

Of the waves, Slater says, "It looked 15 to 20 feet, which was manageable—with the occasional 30-footer. From the channel it really did look doable."

Slater, like Skindog, was measuring by the macho Hawaiian big wave surfer's scale; judging the simple top-to-bottom face height, he meant the waves looked 25 to 30 feet, with the occasional 40-footer. Despite what would happen a little while on, it's always been surprising to me that Slater and Walla's paddle into the unknown isn't a more celebrated event. Partly this has to do with the fact that all the focus at the time was on towsurfing, and also, in later written accounts, the modest Slater would downplay what actually happened. Yet it's fair to say that, up to that day and at least until January 2008, Evan Slater and

John Walla committed either the bravest or the stupidest act in the long, storied history of big wave surfing. Probably both.

"There's one thing I wanna say," says Slater in his defense today. "What we did was probably pretty dumb. It wasn't meant to be any publicity stunt or getting attention kind of thing. Looking at the waves from the channel from the boat, it honestly—it was pretty hard to tell how big it was."

Evan leapt into the deep water, clutching his nine-foot-eight Rusty, a big, stable paddle machine you might use for the biggest Hawaiian outer reefs. Walla followed atop a 10-foot Rich Hynson. About a quarter mile of rapidly shoaling water separated them from the surf zone. To Slater the water felt thick, foreign. "It was like the stuff they add to cling peaches," he says. "I think it has to do with this idea that the ocean is different, denser, when it's so cold and you're a hundred miles offshore."

They immediately noticed that, from eye level, the swells were steaming in incredibly fast.

The duo took up a wary position off the kelp-tangled western edge of Larry's Bowl, alongside a baffled sea lion. There they could watch the waves from fairly close to see how they were actually hitting the reef. A medium set came through. The roar was terrific. Evil boils churned.

"We're really whistling in the dark on this one," Walla said to Slater.

Evan agreed. He says, "Once you got out there, you realized, not only how difficult it was to actually get in a spot to catch one of those waves, but how vulnerable you were."

The towsurfers were a quarter to a half mile up the point. If something serious went down, and someone on a ski didn't happen to see the paddlers—which was the likely scenario, since none of the towsurfers even knew Slater and Walla were paddling out—there was no hope of rescue by boat. Essentially and for all practical purposes, Slater and Walla had just trekked alone into an utter wilderness. They couldn't tell where to line up, there were no points of reference, there was too much current, and the waves were far shiftier than Todos Santos or Maverick's, seemingly bending in and almost approaching at a sideways angle compared to the North Peak—just like Sunset. Then there was another factor, not noticeable from the boat—four-foot-tall refraction waves were marching across the big waves at a perpendicular angle, putting a bizarre bounce on the water. Stroking in was going to be incredibly, incredibly difficult. But when a solid wave, 25 feet on the face, swung toward them, Walla nonetheless started paddling.

"Go, go, go!" Slater shouted.

"I was kind of in it, kind of not," says Walla. "I was like, dude, no. I backed out."

The leading edge of the swell's steeper, twenty-second energy was sweeping the Cortes Bank and another set quickly followed. Walla and Slater paddled over the waves, exhaling through the storms of spray with a whoosh. "Next set, for sure," Slater told Walla.

A few years after the death of Foo, and not long before his horrific knee injury at Maverick's, Evan Slater was nominated for a "Worst Wipeout Award" at the 1997 *Surfer* Magazine Video Awards. He endured a three-story upside-down elevator drop to hell at Maverick's that drew horrified gasps from the crowd and was augured to the bottom. Yet he climbed his leash, found the surface, and paddled right back out to the lineup. From that moment, everyone at *Surfer* regarded Evan Slater as both preternaturally calm and almost certifiably insane. I asked him what he was thinking when he took the drop, and I will never forget his answer.

"I was thinking I was gonna make it," he said. "A big wave's different from a small wave in that you have to start paddling so far before the wave comes. So you can't start paddling with any hesitation. You have to *know* you're going to make it. I know that sounds kind of weird, though, because sometimes you don't."

"Evan always eats shit," Walla says with a laugh. "Every session. He'll get some sick ones, but he's always the guy who's like, air-dropping 40-footers and getting slammed. He's just crazy when it comes to big waves. He's a lunatic."

A lunatic was about to eat shit.

Slater and Walla noticed the next set. Gerlach, Parsons, Mel, and Skindog noticed it, too. It was absolutely huge. These waves seemed to be focusing on *both* the North Peak and Larry's Bowl. Arms dug and throttles roared.

Slater set his sights on the first wave. He paddled like his life depended on it—since it did. Walla echoed his words—"Go! Go! Go!"—as he stroked over the top of the wave in the opposite direction. Slater passed Walla running downhill at better than twenty knots.

"I was hell-bent on trying to catch it," says Slater. "I put everything I had into it."

Slater was close, but not close enough. The wave, eventually, barely rolled beneath him. An old Sunset Beach adage suddenly echoed between his ears. "Never paddle for the first wave of the set." Slater was now thirty, forty yards inside of Walla. With a dawning horror, he looked over his shoulder.

"I paddled over Evan's wave and there was another one," says Walla. "It was the full thing—a seven-, eight-wave set. I was like, 'Oh, this isn't looking good.'"

Each wave was successively bigger—a nightmare stairway to heaven with no end in sight. Walla and Slater broke into a full sprint, trying to clear them before they broke. Slater watched Walla grab the edges of his board and sink it deep into the first wave, a move called a duckdive. He soon followed suit. Both managed to duck the next one, too. But it wasn't enough. The ghosts of Kinkipar understood what was about to happen.

Out at the top of the point, Skindog hurled Peter Mel onto a massive roping wave that might connect all the way through to Larry's Bowl and far beyond. Mel etched a series of fifty-yard-wide half-moon-shaped turns into the wave's face, with Dana Brown's cameraman following him closely. Slater and Walla saw Mel flying toward them. The wave reared up majestically on Larry's Bowl, far outside of the paddling surfers. Mel was astonished to catch a glimpse of Walla far below and inside. Mel had time to think, "What the hell?" There was no way Walla was going to punch through. He ditched his board as a hillside crashed down on top of him. His fifteen-foot leash, a cord capable of supporting well over a hundred pounds, stretched like an Acme slingshot in a Road Runner cartoon and snapped. Walla was a ragdoll, but his experience was not even close to that of Slater. It's not shown, or even mentioned, in the film *Step into Liquid,* but this remains the most terrifying moment of Evan Slater's terrifying life.

"I remember watching John punch through the lip and ditch his board," Evan says. "He was out beyond me, and there was nothing I could do but watch the wave explode. Top to bottom, it was probably a legitimate 40-foot face. [Actually, 50-foot, James Thompson says.] It was a *big wave*. And you know, I've never really panicked before in big waves—never felt like the end was going to happen. I've never felt so on my own, alone, nervous, scared in big waves as I did when that wave broke in front of me. You're a hundred miles out, and there's a hundred different reasons I can understand why I panicked, I guess. I just remember putting up my hand halfway and yelling as the wave's coming. It was just like, 'Heeellpp!'"

"I was praying for him," says Thompson.

The wave enveloped Slater in a shockwave of almost unbelievable force. He doesn't know how far down he was driven, but it was surely very deep. He had no idea that a concrete ship had once met its doom out here. He curled up in a ball and tried to relax as he was blasted southward. Had he been caught in a particularly bad downward shear, skin or his leash could have been easily snagged by reef, rebar, or worse, he might have been forever sandwiched between the seafloor and a broken slab of concrete. Instead, his leash snapped and he was carried far, far inside—perhaps an eighth of a mile underwater. What probably saved him—indeed, what saves many surfers in big wave wipeouts—is the fact that guys like Slater can hold their breath for at least two minutes, and the very water that threatens to dismember you also becomes an aerated cushion. Our bodies are around two-thirds water, so you sort of become one with the molecules. After being driven deep, Slater may have actually risen back up, traveling much of the distance above sea level, spinning and bouncing like a bingo ball in the white water. Or he may have been below. Only God knows for sure. Slater wondered about his life-insurance policy. Was it up to date? Did he have enough coverage? Would he ever see Jennifer and his six-month-old daughter, Peyton, again?

After maybe a half a minute alone with his thoughts, Slater emerged, gasping for air in a deep cappuccino foam that made breathing terribly difficult. Had the froth been any deeper, he would have been unable to inhale anything but bubbles, and he would have suffocated. Two more walls of white water swept over him, but he had now traveled so far that much of their force had dissipated. Slater was dizzy and lightheaded, but he had survived the worst.

A badly shaken John Walla swam over to Rob Brown's catamaran. His board was gone, but he was just happy to be alive.

On their WaveRunners, Laine, Skindog and Mel rescued Slater and went to hunt for the lost boards. Skinny and Mel were pissed. "It would have been nice if they'd come to us and gone, 'What's the lineup?'" says Mel. "But they were on their own mission. 'We're gonna paddle. You guys be the motorheads.'"

"At that point, I really felt like a liability," says Slater. "We're just, you know, creating more hazards. I thought, this is probably a good time to just watch the show."

Incredibly, the surfboards were found, unharmed, and returned to their owners. "I guess this is Towville," Slater told Walla. They climbed back onto their boards and paddled over to the bleeding edge of the breaking waves to watch the greatest spectator event since *Ben-Hur*.

Walla laughs today. "But Evan, he still really, *really* wanted to catch a wave."

Meanwhile, at the top of the point, Parsons and Gerlach were having the time of their lives, oblivious to Slater and Walla's near-death experiences. On one wave, Gerlach was pulled to his feet in time to see a huge shark swimming by, and he then became ensconced in another spinning barrel as big as a cathedral—disappearing so far back that he couldn't see a damned thing. "There's this feeling—that any second the wave is just going to go booooom," he says. "And you're gonna get drilled harder than you've ever been drilled before. You're saying to yourself, 'Make it, make it, don't fall.' If you make it, it's blind luck. It's surfing by Braille."

Gerlach rocketed out, so amped that he thought his head would explode. Mike Parsons had yet to get a chance at the rope. He was twitching like a junkie and about to crawl out of his skin. *What if the wind comes up?* he asked himself. *What if I don't get a chance to surf?*

"Brad, I gotta get a wave!" Parsons shouted.

Not fifteen seconds after exchanging places—Brad on the ski, Mike in the water—both surfers noticed that in the distance, well off Larry's Bowl, a big wave, a *huge* wave, was coming. What made it so much taller than the others is not something that could ever be *definitively* answered, but it's probable that a couple of swells merged as they reached Bishop Rock at the same time. Mike Parsons was looking at a rogue. John Walla and Evan Slater saw it, too. They paddled like hell for very deep water.

"When I saw it, before we even got on it, it was like one of those things," says Gerlach. "I looked over it and I'm like . . . I didn't even say any words. I just nodded my head like, 'You *want* this thing, don't you? Even though you haven't had a wave yet for the day, you're not gonna wave me off. I know you better than that. I just wanted to double-check, right?' He gave me the same look back like, 'Yeah, yeah. Put me *on* it.' So I was like, *okay*. I just made sure I was real . . . you really have to finesse the whole getting off the ski, getting on, grabbing the rope. If you miss the rope, you miss the timing by two seconds, you have to swim to grab it back and it gets really tough. So I just made sure I finessed it."

Gerlach smoothly juiced the throttle to bring Parsons up out of the water and onto his surfboard and mentally calculated his angle of approach. He says, "I'm like, 'Okay, I'm putting him on this thing.' And it was sticking *way, way* up and I said, 'Oh, yeah! The swell's gonna kick in now.'"

Gerlach was stoked for Parsons, but he was also covetous. A historic wave like this only comes along once, and then only if you're lucky, in any surfer's life. *This is how you do it, Snips*, Gerlach said to himself. *You fucking little bastard.*

Parsons let go of the rope, and Gerlach veered off like a fighter pilot. Parsons faded left, quickly losing sight of his old rival. The sound of the ski faded into the distance, and in a few seconds all Parsons could hear was the roaring wind and the sound of his board slightly pattering across butter smooth water. For a brief moment, the sensation was almost . . . *peaceful*. After dropping swiftly for a few seconds, though, Parsons had a disconcerting realization. On most of the waves he had seen ridden so far, you could fade a bit, crouch into a bottom turn hard enough to drain the blood from your brain, and then either line up for the best series of cutbacks of your life or thread a wormhole barrel. This wave simply didn't have a bottom. *Holy shit*, he thought. *This thing's gonna drop forever.*

Parsons angled toward the right and his board began to chatter with the speed. The wave kissed the sky.

Rob Brown was transfixed in horror and fascination. With only thirty-six shots on his camera, he had to consciously remind himself not to blow through his whole roll of film. "I didn't have a comparison for the wave because it was so much bigger—I mean, I couldn't even compare it to anything else," Brown says. "The size and the sound—it was just such a leap over anything I'd *ever* seen. I was sitting there going click, click, click, watching the counter go down frame by frame, telling myself, 'Relax. Mike is gonna die right here, and right now, but you're gonna do your job.'"

Aaron Chang and Randy Laine were directly inside of Parsons and gaped in wonder. In all their years on the water they'd never been granted a view quite like this.

The wave threw forward an enormous lip. The concussion rattled Parsons's brain. He was inches from death. He tried to focus on the boils in his path and concentrate on where the nose of his humming, skittering surfboard was taking him. If the wave, somewhat slowed by the Bishop Rock, was marching over the reef at forty to fifty miles per hour, Parsons was probably doing sixty-plus down its endless face, a moving slope of ocean nearly a football field in length. Maybe Parsons was going to drop forever.

Parsons says, "I was thinking to myself, *get in the right spot*. Then I was just locked in. I mean I *knew* I was riding the biggest wave of my life, but I was just focused on *make it, make it, make it*. No mistake."

He blasted past a cheering Mel and Skindog, and then Slater and Walla—who sat in reverent awe a stone's throw away on the shoulder. Had they still been trying to paddle in, they would have been in the bull's-eye of a wave perhaps 70 feet high, and they would have probably died. "I think he got weeded," Walla yelled through the torrent of spray.

But Parsons stayed planted. To Flame and Sharp in Vince Natali's plane, he appeared positively Lilliputian. Cheers drowned out the propeller.

Parsons kicked off a hundred yards inside of Slater and Walla and the small gallery erupted.

"I've watched and photographed Parsons since he was fourteen," Brown says. "That was the pinnacle of his long surfing career."

Gerlach picked up Parsons, and they throttled back out to Rob Brown on his catamaran. "Guys, that's the pinnacle," Rob said. "We've done it. Stop while you're ahead. Let's go before someone dies."

But no way, no *way*, was Parsons leaving. The next set held another wave—nearly as tall. The skinny kid from San Clemente careened down the spine of another dragon. "I remember going, *Wow, this is a huge one, too,*" Parsons says. "Coming off the bottom, I just felt it breathing down my neck."

This time the dragon's breath burned. Once the wave broke, "it just swatted me," Parsons says. "Blew me forward out of my foot straps, straight down. I remember going, *Oh my God, you're so deep.*"

Parsons followed Evan Slater over the *Jalisco*'s final runway, his eardrums ballooning inward. He had been much farther out than Slater, though, and his hold-down was much longer. Yet unlike Slater, Parsons had the great advantage of a flotation vest. Still, the wave was violent enough to rip a skintight glove right off his hand. Thankfully, his life jacket straps held. Parsons was bouncing along through the column of white water like an avalanche victim—trying to relax and concentrate on which way was up. Stars began to form in his peripheral vision. And then, just like that, he was blown to the surface. Eyes big as saucers, and with another wave bearing down, he turned to see Gerlach racing toward him.

Brad saved Mike, but there was still one casualty. Parsons's favorite tow-board was simply *gone*. They searched up and down the reef on the ski. Natali, Flame, and Sharp tried to find it from the air. Perhaps it had been stuffed into the hull of the *Jalisco*. Perhaps it was swirling in the vast patch of foam. They would never know, though it was seen again.

"A friend of mine who lives here in Santa Cruz named Sam Samson was on a sailboat race to Cabo," says Pete Mel. "He saw Parsons's board out in the middle of the ocean. The first time he'd ever seen a towboard. He goes 'What the hell is that?' Got a good visual of it—yellow logos and yellow rails. He was in a race, though. He couldn't turn around. So he left it."

When it came time for lunch, the surfers and lensmen motored back to *Pacific Quest*. Randy Laine told Chang he wanted to have a look at the top peak. The swell actually seemed to be growing a bit larger, and he wanted to see just what his ski could do. He left everyone and motored a lonely mile up the reef. "I was tripping," Laine says. "And I'm not slamming any of the guys surfing, but the biggest sets of the day actually went unridden. I'd ridden big Avalanches and Todos well over 50 feet, but this was just significantly bigger. There were some rogue frigging sets—I mean, I looked at some, and there were ones I wanted to call 90 feet on the face but you couldn't get a perspective because no one was on them."

When a true monster set came through, Laine lined himself up for the second wave, gunned it, and was soon roaring along on his WaveRunner at better than fifty-five miles an hour, trying to stay ahead of it. "I didn't realize how fast the waves really were," he says. "I had thought previous to this trip that the third reefs in Hawaii had all the speed—because I'd ridden all of those—all of 'em. This moved even faster."

The wave swallowed Laine whole. He braced himself and held the handlebars in a death grip and was bounced around like a piece of popcorn. He had time to think clearly that if he came off the ski, no one would see him and he would simply drown. Eventually, he was somehow blasted out. "It's just a miracle I survived it," he said. "It was a terrifying thing to realize—that even with the fastest ski, you could just not survive."

Parsons and Laine's waves seemed to have broken at the absolute apex of the swell, which afterward gently subsided. The day was gorgeous and the wave faces became increasingly inviting. Skindog kept offering to tow Evan Slater into a few, and eventually the dedicated paddle surfer relented. "It took me a while to get up," says Slater. "But I got a couple of in-betweeners. The speed of the waves was 25, 30 percent faster than anything I'd ever seen. After fifteen minutes Skinny was over it and kicked me off."

"The surfing Pete and Skinny were doing, it was so ridiculous," says James Thompson. "It was also really interesting to see their skill levels. Gerlach and Parsons were a little more conservative, sitting out the back and towing into the bombs. Skinny and Pete, though, they were going straight at 50-footers, flipping the ski around, and whipping the guy into the pocket so he had extra speed. Pete would do a carve up the face and come back into it with another carve."

"These swells were *moving*, you know?" says Pete. "Even the slopey ones, when you're on the face, the g-forces are just kicking when you're doing your turns. It was pristine and blue and kind of *inviting*. It felt so cool. Like no other. It was incredible."

Mel towed Skindog into a 25-footer, a smooth-as-glass halfpipe from the top of the point. The wave wound down toward Larry's Bowl, but rather than offering a makeable barrel, it swallowed Skindog alive and smote him with a three-wave battering that should have led to a lifetime of post-traumatic stress disorder. Sensing a seventeen-second opening, Mel rocketed in to pick him up. "Grab the sled!" he yelled. Skinny obliged and Mel goosed the throttle. But the ski's impeller sucked in foam—the aquatic equivalent of a bad case of wheelspin. "I'm in the mud! I'm in the mud!" Mel yelled. They leapt clear at the moment of impact and cartwheeled. By the grace of God, they popped up right by the ski. The machine fired. The impellor grabbed solid water. They were gone.

"For me, it easily goes down as: Is this a dream?" says Skindog. "Why me? How come I'm so lucky? Some of those waves, in my head I wonder, *did that really happen*? I used to think that 30 foot [or what is actually 50 feet] was as big as it gets. You know, 'cause I believe Maverick's can get as big as the ocean will throw. But just seeing what Cortes did with the swell we got. It's on a higher voltage. It's on steroids. But, you know, there's also the weather factor: Is it gonna be clean? We went out and scored an oasis. We might never get it like that again."

Gerlach's eyes glaze over, lost in the recollection. "Whew," he says. "I wish I was there right now. I think about it so much. I wish I was there."

"I felt weightless," says Parsons. "I told Brad, that was it. I mean, I got *the ride of my life*. It just felt like that moment, that day, on that wave. . . . I kicked out and went, *That's what I've been waiting for all my life*."

Parsons had his Greg Noll moment, but rather than being sated, he hungered for more.

A couple of months later, at the first annual Swell/Surfline Big Wave Awards, photos of the winter season's biggest ridden waves were evaluated by a slate of judges—Bill Sharp, Sean Collins, Sam George, Evan Slater, Brock Little, Flippy Hoffman, Mickey Muñoz, and photo editors Les Walker and Flame. In this raucous, baptismal incarnation of the XXL contest, Parsons's 66-foot, $60,000 wave would defeat entries by Peter Mel, Darryl "Flea" Virostko, Noah Johnson, Jay Moriarity, and Laird Hamilton (the first and only time Laird would enter the contest). Declaring one wave "officially" the biggest on Earth is always a perilous business, especially when you have to take into account all the proud slayers of Jaws mammoths, Maverick's mackers, and blurry Outer Log Cabins behemoths. Nonetheless, the overlords at Guinness felt sufficiently awed to declare Mike's wave a world record.

When the dust had settled, Sharp, Flame, Collins, and Parsons took a hard look at the swell specifics from the day at Cortes. The numbers were fairly staggering. A 15- to 18-foot groundswell had produced a wave 66-feet high on the face. But bigger swells hit the Cortes Bank. Much bigger. On the right swell, the unthinkable was truly possible.

The quest for the 100-foot wave had begun.

Chapter 10:

MUTINY
ON THE
BOUNTY

"Men, this gold is mine, for I earned it; but I shall let it abide
here till the White Whale is dead; and then, whosoever of
ye first raises him, upon the day he shall be killed, this gold
is that man's; and if on that day I shall again raise him,
then, ten times its sum shall be divided among all of ye!
Away now!"

—Ahab, from Herman Melville's *Moby-Dick*, 1851

It was a shot heard round the world.

On January 22, 2001, the venerable *Los Angeles Times* plunked down before 1.5 million readers with news of a remarkable maritime occurrence just one hundred miles offshore. On the front page, adjacent to a story about George W. Bush's first day as president, read the headline: "Surfers Catch Monster Waves Off California." The news was scarcely to be believed. A small team of daredevils had challenged the biggest waves ever encountered in the history of their sport in a location all but unimaginable. Yet if the words seemed sensational, a turn of the page brought proof. Mount Everest had not only been discovered looming just beyond the Hollywood hills, it had been summited the first day.

From his terrifying ringside seat, Dana Brown captured a session no less groundbreaking than the waves his father, Bruce Brown, had filmed a generation earlier at Cape Saint Francis in South Africa for the film *Endless Summer*. Brown and Sharp edited together some footage, and feeds were soon caroming off every TV news network satellite in the sky. It was unprecedented, death-defying, and radiantly newsworthy.

Sharp wondered, if there were occasional 100-footers atop the Cortes Bank, what other freaks of bathymetry and swell lurked just over the Malibu horizon? A "K2" reef set in the bull's-eye of the Roaring Forties? A submarine volcano far up north? An uncharted reef off Midway Island?

His fertile mind imagined a documentary film project with a simple concept: explore the world with the best big wave surfers looking for 100-foot waves.

Sean Collins could provide the forecasting, and the lineup would consist of high-profile members of the big wave fraternity: Mike Parsons, Brad Gerlach, Ken Bradshaw, Brian Keaulana, Peter Mel, Skindog, Brock Little, Shane Dorian, Kelly Slater, and maybe, hopefully, Laird Hamilton, Dave Kalama, and the Maui crew. Since Billabong already sponsored the XXL contest and its thousand-dollar-a-foot annual prize, why not suggest that they offer a *half-million-dollar* reward to the surfer to ride the first documented 100-footer? Billabong wasn't going to let a K2-size opportunity pass them by again, and they agreed. Sharp's first working title was "Project Sea Monster," but he soon settled on something a little more dignified—*The Odyssey*—creating an event and a film that sounded like surfing's first high-dollar Homerian epic.

Sharp laid down his logic in an article in *Transworld Business*, the *Wall Street Journal* for the extreme sports set. "A lot of pro surfing is about trying to make surfing into a jock-strap sport," he said. "*The Odyssey* is Jacques Cousteau meets Evel Knievel meets *Crocodile Hunter* meets *Jackass*. It's not nearly as contrived as having a guy put on a hot pink jersey and try to do forty-seven turns on a 2-foot wave."

Sharp raised an interesting point. Unlike mainstream sports, which are largely spectator- and competition-driven, surfing is fanatically *participant*-driven. Many don't, in fact, consider it a sport at all. That's why a substantial percentage of surfers view competition and the ASP World Tour as contrived and even antithetical to surfing's heart, which should ideally beat in soulful communion between surfer and wave. So even if *The Billabong Odyssey* was based on a somewhat outlandish premise, it was really more of a globetrotting adventure in the mode of *Endless Summer* than a further attempt to turn surfing into a middle-class-friendly professional sport like football or Nascar. The same was true of the XXL Awards. In fact, to prevent some renegade glory hunter from doing something profoundly stupid, Billabong's prize money would only be offered to sixty-four surfers on a predetermined roster. Sharp intended to lionize a cadre of hellmen and perhaps resurrect a few careers in the process. It was not his intention to inspire hordes of glory-seeking yahoos—though that is what he would eventually be accused of.

"We don't know what the limit is," he told *Transworld*. "And that's an amazing thing."

Indeed, the waves that towsurfing made available were blowing minds right and left. The year before the Jet Ski allowed Mike Parsons to descend the *tallest* wave ever ridden, it allowed Laird Hamilton to conquer the *thickest* at a warm-water freak of nature in Tahiti called Teahupoo (pronounced cho-poo; rough translation "Broken Skulls"). At this break, swells sweep out of four-thousand-foot-deep ocean and lumber onto a five-foot-deep reef. The resulting wave doesn't go so much up as *out*. On August 17, 2000, Derrick Doerner slung Laird into a flawlessly glassy warping Teahupoo barrel. Atop a dart dubbed *Excaliber*, and in a flouting of the laws of physics, Hamilton disappeared into a cylindrical abyss of foam and spray before being shotgunned out. The wave was so inhumanly massive and powerful that the experience of riding it reduced Hamilton to tears. *Surfer* magazine carried a single cover line: *Oh My God.*

These were the sorts of ultimate, soul-shaking encounters that surfing, particularly big wave surfing, was all about, and no one exemplified this more than Laird, who was towsurfing's answer to Michael Jordan. His presence would be requisite for Sharp's *Odyssey* project. Sharp met Hamilton and his agent, Jane Kachmer, in Los Angeles, but things didn't go well. First, Hamilton's sponsor, Oxbow clothing, was less than comfortable allowing their star to compete in an event so closely tied with Billabong, a competitor. This corporate anxiety could have perhaps been ironed out, but Sharp says the arrangement fell apart for a far simpler reason: With two of the bigger egos in surfing, each wanted more control over the project than the other was willing to cede.

In the years to come, Sharp and Hamilton would come to disagree over the role of towsurfing, the impact of contests and money—with the XXLs being the most obvious target—and indeed the very morality and purity of purpose that modern, hydrocarbon and horsepower-driven big wave hunting was somehow supposed to represent. Eventually their split over the XXLs in particular would grow into a mutual and at times extremely personal antagonism. Towsurfing was about to transmogrify into a Frankenstein's monster that would threaten to ruin their favorite breaks—Jaws, Todos Santos, and Cortes Bank. Hamilton blamed Sharp for this, and when that happened, Sharp shot right back. Yet in the end, each was probably no more guilty than the other. For after seeing Parsons at Cortes, Hamilton at Teahupoo, and *both* surfers at Jaws, the mongrel hordes were coming whether anyone liked it or not.

Bill Sharp launched the first of just over a year's worth of *Odyssey* missions in late 2001. While Sharp glibly proclaimed the *Odyssey* team the "Delta Force of Surfing," he found that mobilizing those forces—Jet Skis, a quiver of tow and paddle boards, airline tickets, and all the related hardware needed for towsurfing—on a moment's notice to be damned difficult. Further complicating things, despite all its satellite-assisted number-crunching, surf forecasting remained as much art as science.

Still, Sharp's argonauts successfully chased down waves from Oregon to Chile to the South Pacific and had their share of near-death experiences. At a fearsome French beach break, Flea nearly snapped his neck after ducking under an iron curtain, while Sharp was almost decapitated by his drowned Jet Ski. Gerlach simply lost Parsons as daylight waned. "Mike was just gone," Sharp says. "I'm going, 'Oh my God, I just destroyed a six-thousand-dollar ski and killed Mike Parsons.'" Mike endured a relentless pummeling before Gerlach finally spotted his weak waving. Had Parsons not been wearing a life jacket, he surely would have died.

On an aerial recon off western Australia, Sharp spotted an azure mauler with an awful, spitting barrel—Teahupoo's righthand twin. Parsons and Gerlach tentatively sketched into a few rampaging drainpipes that scared them to death. In honor of Homer, they named the one-eyed monster "Cyclops."

When they tracked a forecasted swell at Todos Santos, they ran into John Walla. With all the excitement over Jaws, Maverick's, and Cortes Bank, the focus of big wave surfing had largely shifted away from Todos. "We had been getting perfect glass, ceiling-high barrels to ourselves for years," Walla says. "Then Bill and the *Odyssey* came out there. It wasn't very big, but they wanted to ski. Bill wanted us out of the lineup. They've gotta make money—produce for Billabong. I'm like, 'I *ain't* getting out.' Bill and I got in a huge argument."

Walla was paddle surfing, and for reasons of basic safety and established big wave courtesy, the unwritten rule is that towsurfers generally don't share waves with paddle surfers. If paddlers are out, towsurfers motor off or wait their turn. Yet the *Odyssey* mercenaries were only equipped for towsurfing, and Walla was put out to be asked to cede what he regarded as his priority to the break. He says, "I actually paddled up to Pete Mel and explained the situation. He was supercool. I'm like, 'Hey, you're more than welcome. I've got a board over there, and you guys are welcome to it. You can *paddle* surf all you want. But I gotta draw a line in the sand somewhere, you know?'"

In early January 2002, several Odysseans descended on Maui. Rumor was that the Tow-In World Cup Competition might run at Jaws, and they wanted to be there. While this was officially a "first annual" event, the very first tow-in contest at Jaws happened on December 26, 2000. It was a quickly organized ad-hoc event called the "Peahi Superbowl," and first prize for winners, and Maui locals, Luke Hargreaves and Sierra Emory was pizza and beer. Laird Hamilton told Evan Slater that he and his buddies sat the event out because there was no prize money, and the event wasn't terribly well thought out. "You can't throw any hint of an ASP format toward a tow-in event," Hamilton said. "Everything needs to be considered: wave selection, how you tow your partner in, how you follow him, and the line you take to pick him up. To simply judge the wave would do the sport an injustice and allow the teams to be totally unaccountable for their methods. On top of that, it's just dangerous."

Then the following year, a Brazilian documentary film company, Estudios Mega, put up a whopping $168,000 to sponsor what became the Tow-In World

Cup. The sponsors gave little consideration to Hamilton's ideas, so Hamilton and most of "Team Strapped"—as his fellow cabal of towsurfers called themselves—decided to sit out the 2002 event as well. The snub did not go unnoticed. Not only were Laird and Kalama towsurfing originators, this was their home turf.

"We basically didn't support the fact that they were putting all the money on something that wasn't quite ready for it yet," Team Strapped crewman Rush Randle told me in an interview shortly after the contest. "We started the sport to have fun. . . . There were people looking to say, 'I'm the best big wave surfer in the world. I don't even know how to surf, but I can get towed onto a big wave.'"

Dave Kalama explained to me: "Say you and I go on a surf trip to Indonesia to find the new, greatest place, and we say, 'Come one and come all, it's attainable to everybody.' I don't think we're putting everyone at unnecessary risk to come and surf a perfect 3- to 6-foot wave. But if we find a 60-, 70-foot wave, and we put it in the magazines, then besides, we go, 'Let's have a contest, and we'll put up a hundred grand and just bring everyone down here—people who've never surfed the place.' . . . I mean, if Jeff Clark had been surfing Maverick's for twenty years, and we say, 'We're gonna have a contest, invite fifty guys you don't even know.' I think he'd say, 'That's not too cool to me.' It wouldn't be responsible. It would be dangerous."

Perhaps not surprisingly, even though neither Sharp nor *The Odyssey* were connected to the contest, Team Strapped refused to even be interviewed for Sharp's documentary (with the notable exception of Randle). Sharp would describe the history of towsurfing in his film without a peep from Laird and Dave Kalama.

Based on Sean Collins's World Cup forecast of big but not apocalyptic waves, Gerlach and Parsons opted not to bring the bigger boards they would have used out at Cortes Bank. But as occasionally happens, Sean's forecast changed. It would be nuking. Despite their woefully inadequate equipment, the siren song of a long-period swell crashing over the reef at Peahi was utterly irresistible.

The pair reached Maliko Gulch, a little nook off the Hana Highway where surfers launch their Jet Skis, to find a full battle zone. Waves were washing clear up into the river, setting ski-towing trucks afloat. Some people were swept into the rocky cliffs. You'd think Mike would have learned a lesson from Maverick's, but he and Brad were unprepared for what they saw. "There was a heaviness out there," says Gerlach. "I was thinking, *shit man, what did they sign me up for?* It was very much like I imagine it would be going into war. So many unknowns. We had never even seen Jaws in person."

They sheepishly asked around in the predawn darkness if they might borrow someone's watercraft. A local fireman named Jay Sniffen offered a tiny, tippy two-seater, half the size and power of the ski they used back home. Chasing down a big wave with it would be damned difficult. "We called it the Hamster," says Gerlach. "We're feeding it carrots and its exercise wheel is our motor."

After Parsons threaded the Hamster through a terrifyingly small hole in the surge and out into the open ocean, they puttered north for twenty minutes, arguing over where, exactly, they were heading. They finally found a small fleet of boats and skis bobbing alongside enormous rooster tails of spray. The contest had yet to officially begin, and so noncontestants, like Kalama and Hamilton, were still catching waves.

"We pulled right up," says Gerlach. "I'll never forget this image of Laird powering down the face—boom!—getting air. Then skipping down the face and—boom! —getting air. He looked like he was just saying, '*Fuuuck* you guys. This is *myyyy* spot.' You think about your own home break and how you have it wired. That was Laird. He didn't have a bad attitude, but his confidence and presence was *heavy*. It was just like, 'There he is. There's the King of Jaws.'"

Then they watched another surfer skipping and tumbling down the face of a 50-plus-footer as if atop the horns of a bull. Pure carnage.

The winds were howling straight offshore, streaking and folding small creases into the silvery gray wave faces. The roar was deafening. Gerlach thought they possessed a true monster quality. He nervously contemplated riding one, thinking to himself, "No wonder they call it Jaws."

Later, I asked Hamilton about opting out of the contest. "We rode the waves we wanted and then went in and watched the *experts*," he said, laughing with sarcasm. "We just wanted to watch—to see what all the *professionals* were going to do out at Jaws. I just think anytime you come down someplace, and you have no experience there, you're not just going to step in and show people who *have* been doing it at that place anything tricky. Now if you have been there, learned, practiced, had the experience, put in the time, then yeah, okay, that's a different story. Then you'd be one of the people who understand the spot—and you probably wouldn't have participated in the circus, anyway."

When the horn sounded, Hamilton and his buddies left the water and posted up on the cliff overlooking Jaws, unwittingly removing themselves from a monent that would become among the most iconic in the history of big wave surfing.

Parsons and Gerlach were in the third heat against Team Strapped members Buzzy Kerbox and Michel Larronde. Parsons didn't want a warm-up wave. One wipeout might kill him, so it would be best to get it over with. "You have the pressure of everyone watching," says Gerlach. "Then there's the pressure of— you don't want to kill your best friend. Plus, we had just scored at Cortes, so people are like, 'Let's see if these guys are the real deal.' Eventually, I just said to myself, 'Okay, I've been surfing for almost thirty years. I know how waves work. We're gonna go out here—we're gonna pick one and go.'"

Prior to the contest, they had studied videos of Laird and Kalama intensely. In the water, they dissected Kerbox and Larronde. Where did they line up? How did they track a swell? When did they drop the rope? Heads were lowered while Gerlach said a little prayer. "God, help us out today."

The first wave of a new set loomed. Kerbox and Larronde throttled up and were gone. The wave was big. But Parsons had a feeling, just a weird twitch borne of a lifetime of instinct, that lurking behind was something bigger. The Hamster was too slow to simply chase a wave down from behind, so when they saw the first beastly wave, Gerlach fed it a carrot and angled Parsons to intersect it.

Gerlach curves and closes his fingers one by one, mimicking the curl of a breaking wave, and sweeps his hand in a horizontal line. "The wave goes like this," he says. "It moves with the bottom—follows the shallows. When you're driving in, the focus is *intense*. Don't put him all the way over on the shoulder, but don't put him in too deep so he can't escape. Eventually, I put him in what I thought was the right spot and let him off."

Suddenly the wave lunged upward and Parsons was *hauling ass*—as fast as he had gone at Cortes, maybe faster. But because the wave was sucking water backward across the reef, it gave the strange sensation of running down an up escalator. Parsons bent his knees to absorb the bumps and simply rocketed straight on, a herd of angry water buffalo nipping at his heels. He narrowly kicked out on the shoulder. Gerlach, unable to keep up, had seen nothing but the back of the wave.

"Dude, I barely survived that one," a hyperventilating Parsons told Gerlach. "Put me on the shoulder. I'm not kidding, *on the shoulder*."

Catching the wave on its less-steep shoulder might make the ride a little less critical, but at least Parsons would be closer to the exit. He cared less about winning and more about surviving. Overhead, helicopter pilot Don Shearer saw Parsons and Gerlach gunning it for another *huge* wave, a *tsunami*. He swung into a near sea-level position while cameraman Peter Fuszard, filming for the contest, zoomed in close enough to see the whites of their eyes.

Gerlach would try the same technique. He had just enough of an angle to get the Hamster over the speeding wave's hump so he could whip Mike in. "I was just like, okay, this is a big, big, big wave, don't put him in too deep," says Gerlach. "I look back to see if he's back there and he's *already let go*. There was no way, no way he should have already let go. I said, 'dude, what are you fucking *doing* man? Fuck. Oh, you *fucked up*.' Then I just look back at him one more time and go, 'well, fucking good luck, man. Dig deep on that talent, buddy, and make it happen.'"

Parsons made a few tight S-turns and bent his knees deep to absorb the bumps, his board blipping a dotted line of wake as he aired over small pieces of chop. He began a brief fade back toward the left, mirroring his actions at Cortes, but realized this was not the place to fade and corrected with a quick jog back to the right.

As the wave began to stand toward vertical, swirling boils of sand were swept up into its face. The wave itself was moving at forty to fifty miles an hour and pushing directly into an offshore wind of ten to fifteen miles an hour. This whipped up an instantaneous gale that pushed against Parsons like a big hand and ratcheted up the chop. "I'm thinking, *Everything's wrong here. This is fuck-ing it,*" Parsons says. "Get to the bottom—just get down as far as you can. Then when I saw it bend at me up ahead, I just figured, *Well, you're really done. You've just got to go as far as you can go.*"

A lifetime in heavy water and hours studying Laird and Kalama told Parsons that his only hope was to set his edge and force a high, fast line. He would attain maximum escape velocity from a critical position right at the bleeding edge where the wave transitioned to vertical. But the bumps and chatter were so fierce that he couldn't change course. His feet were slipping and shifting. He drifted downward. Then at the wave's base, he made a quick, conscious decision. Turn, *hard.* He leaned low, laying every bit of strength his 160 pounds could muster into fins and rail.

He began to turn, but at the same moment, his board unexpectedly slowed as if a hand brake had been yanked. Water had literally begun to boil in the low-pressure lee of his aluminum fins—a strange condition called cavitation. This instantaneous liquid-to-gas transition causes a sound-barrier-like shockwave along a fin's trailing edge that in turn generates enormous drag. Had Parsons not been strapped in, he would have cartwheeled forward. He crouched deep and planed his hand on the water's surface for balance. "I was just like, *you gotta make this. Hold on.*"

As Parsons slowed and was pulled toward the maw, the boiling suddenly abated and he began to drift up the wave's face and accelerate again—very, very rapidly. He was now aimed down the line at a perfect right angle.

"I remember pulling up there and thinking, *Oh my God, what are you doing? You're pulling into the barrel at Jaws.*"

The wave folded over Parsons, enshrining him in a tube about as big around as a Boeing 727. Very few humans have ever stood amid such force and survived. The eruption of mist and spray became so violent that he couldn't see. The blast of air pressure popped his ears. A tremendous shockwave exploded toward the only opening and Parsons was lifted clear off the surface of the water. "I just said to myself, *Hang on, hang on.* I pointed it and just came flying out.'"

Parsons scored the only perfect ten of the competition. He and Gerlach were in first place. Despite a number of stellar rides, even by a terrified Gerlach, the pair would take second overall, ceding $70,000 to Garrett McNamara and his teammate Rodrigo Resende. Thus, it was the second wave Mike Parsons had ever ridden at Jaws, not the contest result, that came to define the day. Don Shearer has ferried cameramen above countless Jaws waves, but never before or since has a ride been captured in quite such an awe-inspring manner. Bill Sharp wasn't at the contest, but when he saw the film, he was spellbound. "I knew instantly it was the best bit of film ever shot of surfing," he said.

Parsons's now iconic wave became the stupefying opener to the film *Billabong Odyssey*, as well as the most downloaded surfing video in the history of YouTube—with 30-plus million downloads (among the many YouTube iterations, that number is probably quite a bit higher). In fact, the clip, improperly titled "Struck in Tsunami," is among the most highly downloaded video clips from any sport on the entire site, or it would be if it were properly labeled as a sports video. Of the clip's more than ten thousand comments, the most common are assertions that the footage is fake and that the surfer must be Laird Hamilton. Of course, it is neither.

Ten months later, on Halloween 2002, conditions looked good for an Odyssey return to the Cortes Bank. Parsons, Gerlach, and Skindog climbed aboard *Pacific Quest* with a small team of newcomers that included World Champ Kelly Slater and a budding Hawaiian hellman named Shane Dorian.

However, the Halloween mission to Cortes was a bit of a disappointment. The surfers found some solid 30- to 40-footers, but it had been nearly twice as big the year before. Still, it was a good warm-up for what would become another epic big wave event.

On November 26, 2002, Mike Parsons and Brad Gerlach were lured back to giant Jaws. It was the first time they'd ever seen a *real* crowd of towsurfers. Clearly, the increased profile of the sport was having an effect. The amped, hustling ski jockeys were yelling and cutting one another off. The surging water hummed with smoky machines and the type of competition for waves that you expected in 6-foot Lower Trestles.

Parsons thinks this is the reason they eventually chose a macker that Brad had no real chance of making. It was the first of what would seem an endless set, and Gerlach could do nothing but brace for impact. "I started deep breathing," he says, doing a hilarious imitation of his rapid breaths. "All the people up there safe on the cliff are like, 'Holy shit, I don't wanna be that person.' I don't wanna be me right now either."

Gerlach was jackhammered. Despite the long hold down, the life vest would, he hoped, help him return to the surface. In the meantime, should he struggle? Go with it? Just how long could he hold his breath? He tumbled and tumbled, giving himself second by second updates. He thought he had enough air, but his topsy-turvy world soon grew fuzzy around the edges. Suddenly, the boiling abated, and he sensed an opening. He stroked for the surface. The biggest wave he'd ever seen from sea level was bearing down on him. He hyperventilated, filling his bloodstream with fresh oxygen. "Then boom," he says. "And these second by second updates again."

He surfaced again, only to be smashed by a third wave, and Gerlach's emotions shifted to a sense of sheer wonder that he was handling a beating of this magnitude with enough clarity to ponder his own mortality. Without a life-jacket he would already be dead. With it, Jesus, just how much could a human being endure? And where the fuck was Mike, anyway?

Again he surfaced, and a ski bore down. It was Kelly Slater. The world champ held out his arm. Brad missed, grabbing the rope instead. Kelly gunned the ski, pulling Brad away and nearly drowning him all over again, churning liters of seawater through his sinuses. But it wasn't far enough. When another wave came, Kelly roared off. Brad was again alone. When he finally surfaced, Mike was waiting. Brad was so keyed up, his blood sparked and crackled so hard with endorphins, that everything seemed overexposed, *white*.

When they returned to the lineup, Laird Hamilton and Dave Kalama motored over. "Whooo, she gave you a little kick in the ass, eh?" Laird said. "She went easy on you then, eh? You're back out here. Good."

This afternoon at Jaws came to be called the "Day of Days." Laird Hamilton was typically insane, bombing and side-slipping down monster waves at breath-taking speeds and catching huge air as easily as you might ollie a snowboard. Australian Ross Clark Jones drove through the biggest barrel Gerlach and Parsons had ever seen negotiated, that is, until Garrett McNamara then ducked into a collapsing mountain that would earn him $5,000 for the XXL's Barrel of the Year. An eighteen-year-old Oahu kid named Makua Rothman was pulled onto the XXL's Wave of the Year, a $66,000 66-footer that would, for the time being, tie Parsons's Cortes world record. Gerlach was judged to have endured the year's worst wipeout—his life being thus valued at a mere $5,000.

The super session demonstrated why there weren't many places in the world like Jaws. For accessibility, only Maverick's was equivalent, and even if it held more potential, Cortes Bank was, logistically speaking, several orders of magnitude more difficult. Yet the day also served notice that a shocking number of surfers were, not only willing to risk everything for a wave, but able to survive wipeouts that, on their face, appeared utterly fatal. If Hamilton and his Team

Strapped friends had cracked open the towsurfing door in the mid-1990s, by winter 2002 it had been blown off its hinges. But the bottom line was, there were only so many giant waves to go around. Towsurfing contests, the regular dismantling of world records, and the resulting high-profile media attention were fueling an almost self-destructively successful interest in the sport. If any surfer doubted this, they only had to attend raucous NOAA meetings where angry environmentalists were demanding bans on towsurfing from San Francisco to Santa Cruz—or they only had to wait until the winter of 2003–04, when the circus discovered and descended on the Cortes Bank, clowns and all.

Expeditions to Cortes Bank would soon start to become regular occurrences, yet one notable absentee was the break's Christopher Columbus, Larry "Flame" Moore. In fact, after leading the charge in January 2001, Flame would never venture out to Cortes Bank again.

The years following the Bank's revelation were tough for Flame. Most of his adult life had been defined by *Surfing* magazine and a *Groundhog Day*-like routine of surf, work, and family, but then he left *Surfing* in 2000 to help launch the ill-fated Web site Swell.com. Swell chewed through untold millions of venture capital dollars and went flat broke during the dot-com crash of 2001. Swell had initially paid Sean Collins millions for Surfline, but the savvy Collins eventually bought his brainchild back, along with all of Swell—for pennies on the dollar. Flame returned to *Surfing* with a job intact, but his longstanding professional relationships were strained. He lost considerable weight, his walking gait changed, and he was suddenly looking older. Stress and politics were wearing him out. He called his wife, Candace, one Saturday in late 2002 and said, "You gotta come home. I'm falling apart."

Candace called fellow Cortes Bank pioneer George Hulse, who was still shepherding San Clemente's growing Shoreline Church. Hulse's one-time mentor now leaned on his young disciple. They talked for a long time. "I'd just taken a class on burnout," says Hulse. "Larry was just going down the list and reciting the classic—headaches, fatigue, having a hard time staying focused. We prayed on it and I gave him a book. I think it was called *Beating Burnout*."

For a while things seemed to improve.

"But he had told me that his headaches had been getting pretty severe," says Hulse. "I told him, 'This could be something physiological, and you need to get it checked out.' Of course, we weren't even aware of how bad things were going to get."

Only a few weeks after he and George Hulse first talked, "it really all fell apart," says Candace. "He'd get lost at Costco. He had this olfactory thing. When he went to work, he thought he smelled petroleum. One day he drove up on the grass and got escorted home by the police."

On December 31, 2002, Candace and her mother checked Flame into a Mission Viejo emergency room. A brain scan revealed what Candace had feared. Larry's skull now shared its space with an aggressive tumor—a grade-four blastoma. Even with emergency surgery, radiation, and chemo, Flame probably had no more than fifteen months to live. Still, surgery was performed, leaving a stunning scar along the right side of his skull, while chemotherapy dripped poison into his veins. The diagnosis and treatment were, of course, hell on Flame and Candace, but it perhaps most profoundly affected their son, Colin, who was not only in his tough middle teenage years but had been deaf since birth. Flame was stripped of his fine motor skills to the degree that it became very tough for him to use sign language. "The burden of parenting really fell on me long before Larry died," Candace says.

However, for the moment and for the next couple of years, Flame battled with his typical intensity, vastly exceeding his doctor's expectations and even returning to work at *Surfing* as much as he could. Flame had documented the rise of a generation of big wave surfers who would come to change the very definition of what was possible on a surfboard. His illness, though, coincided with the sudden rise of a young man who would come to eclipse anything even Flame had ever seen.

While more and more surfers were experimenting with horsepower, none would come to have a greater impact on the future of big wave surfing than Greg Long, who picked up his first tow rope in winter 2003. Long is that rare combination of seriousness, drive, and thrill-seeking guts—a younger version of and a torchbearer for Mike Parsons, his mentor. But Long has also enjoyed the serendipity of being in the right place at the right time: He was lucky enough to catch one of 2003's most epic waves, and his Kodachrome moment became yet another reason for what happened next at Cortes Bank. Then as the decade unfolded, Long piled up the awards, records, and accolades, while leading the charge to return big wave surfing to its woolliest paddle surfing roots. He would also come the closest anyone ever has to catching and riding Flame's legendary 100-foot Cortes monster.

Greg was born in 1983 in San Clemente, California, and his father, Steve, was the head lifeguard for all of San Onofre State Beach. Steve and his wife, Jan, raised their three children in a fifty-year-old wood-framed cottage on the grounds of San Clemente State Park. Greg's older sister, Heather, was born in 1980, and Rusty, his older brother by two years, was named in honor of a classy

Encinitas soul surfer named Rusty Miller. Steve Long named his second son after a big wave surfer he respected immensely, the legendary Greg Noll.

Steve planted his kids on the nose of a longboard at San Onofre before they could walk and was hauling them down to Baja to camp by toddlerhood. Rusty and Greg were soon bombing the hills above Calafia Beach aboard skateboards and BMX bikes and ducking into the strand's thumping beach break atop boogie boards. The boys made good grades and became talented soccer players and a seemingly psychic little league pitcher/catcher duo. Heather became a stunning, fearless waterwoman.

By middle school, the kids were experienced free divers who knew more about the local hiding spots for lobster and corvina than the crustiest old long-boarders at San Onofre Point. They also came to know, and see, the consequences of Dad's work firsthand. The ocean can kill you. Plan accordingly.

The Longs learned to surf all of the waves that peel from San Onofre through Trestles: Church's, Middles, Uppers, Barbwires, and Cottons Point, but the high-performance right and left peaks at Lower Trestles became the brothers' specialty. Even on days when it seemed you could walk atop the heads of surfers to the lineup, the lean, swarthy young duo learned to slay their share at Lowers. Rusty was a quiet young Zen warrior. Greg was precocious beyond his years, regularly butting heads with guys—and girls—three times his age. "God, he had such a mouth," Rusty laughs.

By the mid-1990s, San Clemente was shedding its image as a somewhat sleepy surf town. In 1992, *Esquire* magazine ran a cover story on the life aquatic of Herbie, Dibi, Christian, and Nathan Fletcher, and other locals ascended to national recognition: Shane Beschen, Chris Ward, and the big wave–charging brothers McNulty. *What's Really Goin' On*—released by a small homegrown surf company aptly named . . . lost—captured a hilariously and disturbingly dysfunctional surf town, portraying San Clemente as a barrio capital for hard partying new school surf punks. It was a world of drug-and-booze-fueled *Jackass* pranks, with kids pushing each other down a deadly hill in a barrel or lighting a buddy's hair on fire with an aerosol can for laughs. But the Longs were not among the donkeys. "We couldn't get away with anything," says Rusty. "We had Dad and all the lifeguards around."

Sounding like his boyhood idol Mike Parsons, Greg says, "Our parents were very open about partying, drugs, alcohol, and using recreationally, and how for some, it's part of their everyday lifestyle. That was never a point for me. I wanted to be a professional surfer. I didn't want anything to slow me down.'"

Rusty probably took his first puff of weed at around fifteen. "I've used it for yoga," he says, "stretching deeper. It slows the thought process a bit—helps with my flow. But I'm like my brother in that I never had the urge to get into anything else that was going to get in the way of my goals."

Also like Parsons, Rusty and Greg both gravitated toward the biggest waves they could find. Why they were so attracted to big waves, becoming classic high-sensation seekers themselves, is impossible for them to say; when they grew up, San Clemente was rife with surfers, yet few chose the path they did. Perhaps acquiring a taste for the thrill of surfing in infancy had something to do with it. Perhaps having a lifeguard father whose profession involved an adrenaline- and wave-filled life did, too. In any case, both can pinpoint the moments that set the trajectory of their adult lives.

When Rusty was fifteen, he pulled an airdrop on a Sunset Beach macker. "That one wave set the hook," he says. During the stormy El Niño winter of 1997–98, Greg began soloing out at San Clemente's West Reef, a lonely bombora that breaks nearly a mile out. "That Big Wednesday swell—the one where [Ken] Bradshaw caught his giant wave at Log Cabins—I sat out on West Reef the entire day," Greg says. "Legitimate 20-foot faces. Biggest I'd ever surfed to that point. I fell on a couple and went through the motions, but I found it was really no worse than bodysurfing at State Park when it's big and closed out. Break your leash, swim a half mile, paddle back out. That's when the spark went off."

Greg recalls how he and Rusty would skateboard across the I-5 to the San Onofre Surf Shop to listen wide eyed as Joe McNulty spun yarns on hairball trips out to confront Todos Santos with Mike Parsons, Evan Slater, and John Walla. Then, on a chilly winter's afternoon in 1999, fifteen-year-old Greg rang up Walla. "Ummm, hi, Johnny," he said. "There's a good northwest swell in the forecast. I was hoping you'd maybe take us down to Todos and show us the ropes."

John Walla had known the Longs for years, and he was even dating Greg's sister, Heather, at the time. Walla well remembers the early winter morning that he left with the Long clan—Steve, Heather, Rusty, and Greg—for Todos Santos. They hired a *panga* out of Ensenada and reached the lineup by Southern California's rush hour. Heather's presence amazed the *panga* captains. They'd never seen a girl out there.

The swell was delivering waves up to 20 feet on the face—lumpy and irregular with the high tide—but still more powerful than West Reef, Sunset, or anything they had ever surfed. "I was amazed at how much water was moving," says Greg. "You can actually hear the boulders clicking and crackling underneath you when you're duckdiving through a wave. I sat next to the rocks watching where John and the other guys were sitting, but I couldn't figure out how to catch a wave. When you'd think it was going to stand up and break, it wouldn't. It would move way past you. You'd have to sit waaayyy deeper, basically sitting underneath it and then paddle as hard as you ever have to catch it."

Eventually, Walla helped coax Heather into a solid wave. Greg started to get a flow and then lined up for a big one. "It was right when Kelly Slater had started doing these carving 360s," Walla says. "Greg drops in on a 20-footer and does a carving 360 off the top. I'm like, 'My God, what's going on out here?'"

Walla knew Greg was a solid surfer, but even good big wave surfers don't pull small wave maneuvers at 20-foot Todos. In fact, it's unheard of.

Late in the day, the biggest wave the Longs had ever seen in person tripped over the reef. "My brother got caught, but me and my dad took the thing *directly* on the head," Greg says. "Just got blasted, came up, and there was another one after that."

The pummeling tore Steve's rotator cuff. "We eventually come up, and he goes, 'You all right?'" says Greg. "I'm fine. I was winded and a little freaked out but I was *psyched.* When you get blasted, in a twisted way, you get this self-gratification and a sense of accomplishment in getting your ass handed to you and walking away from it. It's bizarre. People are like, 'Oh my gosh. If that happened to me once, I'd quit.' But it happens, you survive, and it's like, *Let's do that again.* All I could think about was going back and surfing it—again and again and again."

The brothers became addicted to the rush, yet few were earning a living among big waves. Rusty wasn't interested in competing, and so he set his sights on college. Greg figured that making a hard run at becoming a pro would open the doors to a surf-based livelihood. He trained relentlessly and won the Men's Open Division at the 2001 NSSA Nationals at Lowers—the equivalent of moving from the minor to the major leagues in a day. Dick Baker, the president of Ocean Pacific, offered Greg a sponsorship straight out of high school. The natural expectation would have been that Greg would follow in the footsteps of Mike Parsons, Brad Gerlach, and Kelly Slater—and make his own run at the ASP World Title. But that wasn't where his heart lay. He told Baker, "I really love the adventure side of surfing. I want to go on trips to obscure locations where people haven't surfed before—spend a couple of months riding big waves and immersing myself in the cultures, rather than just going from contest to contest."

To Greg's amazement, Baker replied, "You know what? Do what you're gonna do."

"That was it," says Greg. "I basically had a license for creativity when I was eighteen years old—to go travel and surf where I wanted and make stories happen for the magazines. That's when this big wave thing was taken to a whole new level."

Two years later, in mid-2003, Greg and Rusty made their first journey to South Africa in an "endless winter" pursuit of cold-wave adrenaline. In Capetown, they made the acquaintance of a funny thirty-year-old maniac named Grant "Twiggy" Baker and his bawdy, beautiful girlfriend, Kate Lovemore.

During his formative years, Twiggy roamed the world with his father, Vincent, a rabid fisherman and professional golfer. He cultivated a love of heavy waves while cocooned in the sand-spitting barrels of his native Durban, a thousand miles north of Capetown. Then when Twig was seventeen, Vincent was stabbed to death in a carjacking. Even today, it's not a subject he cares to broach, though it informs his surfing life. Kate says, "It's just so heavy. I've said to him, after some of the things he's done, some of the waves he's ridden, 'Don't you wish your dad was here?' He'll just sort of brush the subject under the carpet."

It took years for Twiggy to fully emerge from a fog of depression. When he did, he forged and tempered nerves of titanium at a lonely, white shark–infested hellhole that Capetown locals called Dungeons. The wave is basically a toothier, shiftier version of Maverick's, fronted by a landscape equal parts Grand Canyon and Big Sur. He also surfed deep in South Africa's desolate Transkei region. To keep from being torn to pieces he wore a great white–repelling "Shark Shield"— a battery-powered surfboard leash that delivers jolts of electricity into the surrounding water.

Greg and Rusty were quickly woven into Capetown's tight-knit circle of Dungeon masters, and despite an age difference of ten years, Greg and Twiggy found they had the same goal, the same mindset—simply paddling and towing into the biggest damn waves they could find, no matter what, no matter when, no matter whether they could afford it. Greg would become Twiggy's California connection, introducing him to guys like Parsons, Mark Renneker, Grant Washburn, Skindog, Peter Mel, and Jeff Clark.

"We started having such a blast, traveling the world for waves," Twiggy says. "It was like, fuckin' *let's go*. We were spending eight months a year surfing. More time together than we were spending with our girlfriends."

Kate laughs and nods. "That's true."

In June of 2003, the Red Bull Big Wave Africa went off in stellar, bonecrushing conditions. The twenty-year-old Californian won despite being the youngest invitee in contest history, while Twiggy finished fourth. For both, the contest marked the start of the most phenomenal run in the history of big wave surfing. A run due in no small part to their friendship.

"Twig and I, we just *clicked*," says Greg. "We've always kind of had two different approaches, but they complemented each other. It's like with my brother. People will say, well, you guys are brothers, you make perfect surfing and towsurfing partners. But no, our personalities really conflict. Rusty's just supermellow, like, 'Take it easy, Greg.'"

Yet, with Twiggy, Greg is the quieter one. Despite being younger, he's more subdued, even if it's a difference of degrees, not nature. With mock seriousness, Greg says, "Twig's the yin to my yang. The peanut butter to my jelly."

Today, Greg Long's name already comes up when surfers argue over who deserves the title of best big wave rider in history. It's a short list, and in recent years Long received the benediction of one of the few other surfers on it, his namesake Greg Noll. The two men, arguably the best big wave surfers of their generations, have become fast friends.

"I was an adrenaline junkie, a glassy-eyed bird dog when the surf came up," Noll says. "And that's what's so great about Greg. He paddles into these big fuckin' waves that everyone else is towing into. I don't know anyone who is doing it as well or as dramatically as he does. I mean, I've tried to understand how he gets away with this shit. You take your lineups, you check your reefs during the slack periods, and then you try to hit 'em on the button when the surf comes. That's what I tried to do, and I still couldn't do what this kid is doing. He has some kind of communication with God or the supernatural. But then what I think is so special, besides his ability, is his attitude. I've never seen a wave rider with such a relaxed, bitchin' attitude."

Long is characteristically humbled to hear Noll's comments. But Noll also puts his finger on an apparent dichotomy in Greg's personality. Is it possible for someone to have "such a relaxed, bitchin' attitude" and also be a glassy-eyed adrenaline junkie, a high-sensation seeker? Long thinks of himself in the Mike Parsons mold, as a very calculating surfer who takes off only on a wave he feels he has a good chance of making—"as opposed to thrill seekers," he says, "who will take a risk even though they're going to get murdered. I mean, I've been on surf trips with people who, if they don't get waves in a week or two, will have to do an extreme act—like swimming a gale-force ocean—just to get the rush they need. Garrett McNamara is probably as extreme a high-sensation seeker as there is. I've seen him take horrendous beatings and just come back laughing. I'm not like that. I never like to fall.

"But still, I know that adrenaline and dopamine rush is where a lot of the satisfaction I get from a big wave session comes from. But there are a lot of other activities where you could get a big rush without such heavy consequences. I've never characterized my need as that. It's the whole adventure side of things, going into the unknown. Seeing a new corner of the world. I mean, if you can't find satisfaction in just being around the ocean and partaking in all the things it has to offer, you're really selling yourself short."

Greg and Rusty Long joined the towsurfing flock after their mid-2003 South African baptism with the specific intention of surfing the Cortes Bank. Steve Long gave them the Jet Ski training materials used by the city of Honolulu, insisting that they learn everything about one-man rescues, slingshots, and rollovers. The boys began training hard along the long, empty stretch of beach south of San Onofre. They were ambivalent about the ski at best, particularly after December 11. On that day, Greg watched an angry paddle surfer named Keith Head get the crap beat out of him at Todos Santos after Head cut the anchor line to a towsurfer's ski.

"A fundamental personality trait of mine is simplicity," Rusty said. "I don't like dealing with this big, inanimate object that you have to fill with gas and oil. But out at Cortes it would be the only way."

By the second week of December, after a year of tantalizing, frustrating waiting and praying for conditions to come together for another Bank job, Mike Parsons and Sean Collins noticed a lump of low pressure that had spun southward off the Aleutian Islands. Initially Parsons had simply thought he'd be heading for Maverick's. Then Rob Brown called. He had just bought a twenty-nine foot Worldcat with twin 250s. When Parsons mentioned Maverick's, Brown said, "Well, have fun."

Parsons noted Brown's sarcasm and asked what his problem was. "I don't care what you're doing," Brown said. "But what the hell do you think I bought this boat for?"

"So what, you want to go to *Cortes*?" asked Parsons.

"I *am* going to Cortes," replied Brown.

Parsons told Brad Gerlach, "Keep your fricking mouth shut on this one." The only people Parsons called were Greg and Rusty Long. They were the hardest young chargers California had produced in some time, and their dad was a lifeguard. Steve Long would come, too, and he lined up longtime coworker and Todos surf veteran Jeff Kramer for rescue. They would journey in a separate boat piloted by Bob Harrington, a paraplegic former surfer who liked to live vicariously through Steve's kids.

A solid, long-period swell at 10 to 14 feet rolled beneath an oily smooth ocean. Cell phones lit like Christmas trees. Everyone wanted to know Mike and Brad's plans for December 17. Lies were told—even to old friends. "Everyone was calling and leaving messages," Mike said. "'Where you going? What are you doing? You ass. Did you already leave? C'mon call me back.' Rody [Peter Mel's new towsurfing partner Adam Repogle] must have called me twenty times at least, and it just killed me to see his name popping up on my phone. Then it was Bill Sharp, then Chris Malloy. The guilt was just eating me away."

That day, when Harrington rounded San Clemente Island, the butterflies in Greg's and Rusty's stomachs morphed into dragonflies. "It was a real, deep glassy, long-interval swell," said Greg. "Stretching from one side of the Pacific to the other."

Fifteen miles out, he and Rusty spotted the Cortes Bank's first waves, which reminded them of moving, snow-capped mountains.

The two teams were all alone. Not a breath of wind. The Longs were dazzled. Some of the waves tumbled and thundered in from far off the top of the point. Others loaded up on the slab of reef Parsons identified as Larry's Bowl and "went square," a surfer term for a wave that throws out a steep, slabbing barrel. Rusty was a little sickened. They looked like sheer cliffs. The only way out of a wave like that would be through the tornado. In their nascent towsurfing careers, they hadn't ventured into anything *remotely* this big, deadly, and perfect.

The brothers figured they'd just watch the pros for a little while. Parsons and Gerlach wanted to explore the top peak. It was so consistent, so glassy—even smoother than 2001. Quiet. Deafening. Surreal. Gorgeous. No airplanes. No film crews. No walkie-talkies. No egos. They glided across edenic watery plains. Few surfers ever stumbled upon conditions like this, and none of these riders might ever find them again. A pair of waves stood majestic over the outer reef, their blue faces unscarred by ski or surfboard tracks. They barreled all the way through to the inside, a distance of better than a half mile. A divine reward for lying. "It was awesome," says Gerlach. "Heaven on Earth."

"There's seriously no way I can put into words what it's like to be able to drop into a perfectly glassy 40-foot wave seemingly without a worry in the world—it's an incredible feeling," added Parsons. "It's like the kind of thing we could only dream about as kids. The only time it sinks in that you're *human* is when you kick out and you're sitting out there floating in the whole scene. There are these giant waves lurking nearby and there's absolutely no sign of land anywhere. You start to feel like you've just been dropped off on Mars without your oxygen tanks."

The Longs started out by methodically surveying the sea bottom and gingerly motoring around the edge of MacRae's Rock, trying to get a sense for where and how the waves were breaking. When they did decide to tow, they took equally cautious strapped steps, pinning down the easier prey like a pair of young jungle cats on their first real hunt. They rode the last waves in the sets so they wouldn't be caught inside. Sliding down the faces at forty-five miles an hour was like skateboarding on a cushion of air. They made no mistakes. "They were on fire," said Parsons. "It was their coming out party."

A perfect, 6-foot wave at Lower Trestles might carry you a hundred yards and last fifteen seconds. A monster at Maverick's might end after a quarter mile, a twenty-second ride. Yet several of Parson and Gerlach's rides lasted the better part of a minute. On one wave, Gerlach counted six fifty-yard-wide silky bottom turns. At 50 feet tall, these were perhaps the most perfect giant point break waves any human had ever surfed. Steve Long, Jeff Kramer, and Bob Harrington watched the whole scene in awe, unaware their boat was sinking.

The sun lowered to the horizon. The boys were spent, but the photo shoots, the weather, the waves were still *so perfect*. Rob Brown called Sean Collins on his satellite phone. "Spend the night," Collins said. "There will still be waves in the morning."

As they stowed gear, Steve realized that the bilge pumps aboard Harrington's boat had been working overtime. He lifted the hatch cover to find an engine bay half full of ocean. Rusty and Greg clambered aboard Rob's boat. Steve and Harrington motored off into the dark to return to shore. They had life jackets and wetsuits, and they alerted the Coast Guard to their situation. Still, it was a terribly spooky journey. They reached the mainland at 2 A.M. to discover that a simple valve had been left open in the live-fish bait tank.

Rob Brown drifted southward off the Bank and eventually into the open ocean. "We all had a good dinner and just kicked it for a little bit and then fell asleep," says Gerlach. "I love sleeping under the stars. There's no light pollution out there, beautiful stargazing. It never felt unsafe or anything. It just felt *right*."

He wondered if he would ever see a day like this again.

Rusty gazed into the deep, black abyss. Whales sang. A weird light suddenly bathed the sea off the distant Bishop Rock. A submarine? A diver? A ghost?

Rob returned to the grave of the *Jalisco* the next morning. They were again all alone. The waves were even bigger.

Two swells pulsed in the water—the dominant, long-period westerly and a secondary northwesterly with a shorter period. Occasionally, when the two swells reached the Bishop stair step at roughly the same time, they turned into rogues.

Rusty was in position for the first of these barn burners. It didn't look all that big to start out, but when it slammed into the top of the shelf off Larry's Bowl, it threw out a wondrous barrel that had the shape of an inverted horseshoe. Rusty's 6-foot 6-inch Timmy Patterson turned squirrelly—like the speed wobbles you might get atop a skateboard. The wave clamped down ahead of him. He rocketed out into daylight.

Aboard the boat, Rob Brown reloaded his Canon with film while his buddy Jon Beachamp steadied his video camera. The *Rolling Stones* blared over the Worldcat's stereo. Brad drove Mike back out to the lineup, passing the brothers as a mammoth reared up. He shouted to Greg, "Go, go, *go!*"

As they disappeared, Gerlach turned to Parsons and said, "Well, that was the wave of the winter."

The brothers had switched roles. Greg plunged six stories straight down on a wave that was arguably as big as Parsons 66-foot bomb of 2001, but even steeper, more hollow. Greg had a moment to briefly contemplate boils and kelp below him. The reef, or maybe the deck of the *Jalisco*, was plainly visible. He felt the g-forces as he turned—the equivalent of bench pressing four hundred, maybe six hundred pounds with his thighs—and fought to hold fast to the wave's epicenter. The wave slammed Larry's Bowl and Greg entered a tornadic barrel, becoming utterly obscured by the spray. He was launched back into the sun.

"The wave of my life," said Greg. "No questions asked."

Greg Long had also just earned his first *Surfer* cover shot. Not bad for your first session at Cortes Bank.

A couple of weeks later, another powerful swell lit up breaks in Hawaii, and hordes of fire breathers descended on Jaws. Greg Long's *Surfer* cover wouldn't hit the newsstands for another month, but a number of Rob Brown's stellar photos had already been leaked by the XXL Web site. With a huge swell steaming toward California, big wave surfers were chomping at the bit. Everyone wanted a piece of the Bishop Rock.

In contrast to the silence that preceded the previous Cortes Bank mission, Sean Collins issued an alert via Surfline, and anybody who had seen the footage at Jaws knew something wicked was bearing down. The fuse was lit.

Invasion forces marshaled, and on January 12, 2004, they attacked. Maverick's founding father Jeff Clark and Grant Washburn joined a crew of kiteboarders and Southern California big wave chargers aboard a seventy-five-foot, jet-powered catamaran called the *Condor Express*. The Longs journeyed with their dad, Jeff Kramer, and Bob Harrington. Rob Brown headed out with Kelly Slater and Ventura surfer Chris Malloy. Bill Sharp lined up an *Odyssey* helicopter to ferry out Parsons, Gerlach, and Hawaiians Shane Dorian and Noah Johnson. Taking a commercial bird that far out wasn't even *legal*.

Another seventy-five-foot boat out of Oceanside called *Electra* held an MTV film crew, along with Garrett McNamara and his tow partner Carlos Burle—a mad Brazilian who had eclipsed Mike Parsons's 66-foot 2001 Cortes wave with a wild 68-footer at Maverick's in 2002. Skindog climbed aboard the *Electra* with his old friend and new towsurfing partner Josh Loya. Flea and Barney joined the party. A small crew of professional freestyle Jet Skiers decided to head out, too.

Yet this swell lost much of its vigor on the passage between Hawaii and the mainland, and the best the Tanner Bank buoy could muster was 9 feet at twenty-second intervals—still good for perhaps triple overhead waves, or about 35 to 40 feet high, but half of what was expected. It was as if Bishop Rock could only muster a relatively meager line of defense against a massive array of firepower.

Overhead, the churning of rotors—a twin-engined airplane, followed by Sharp's helicopter. Sharp was sure he imagined the music of "Flight of the Valkyries" from *Apocalypse Now* playing in the background. The Odysseans leapt from the whirlybird and were chauffeured over to their tiny attack craft. The battle for Waterworld began.

"It was an absolute circus," says Steve Long.

Initial salvos were launched not by surfers but stand-up Jet Skiers, who attempted an attack equal parts Supercross and swamp buggy. They circled the lineup like Apaches attacking a wagon train, laying confused creases and chops in the faces of the waves. Parsons and Gerlach were approached by an overweight man in sweatpants and an Indiana Hoosiers jacket who rode atop what Brad called "an inner tube with a steering wheel." "This is my first time here," he told Brad pointing at his rescue sled. "What is that, you have some kind of pad on the back of your ski there?"

The Hoosiers fan was soon towed straight through the lineup on his knees atop a standard-issue paddle surfboard. "If I see that guy on that inner tube coming down the face," Gerlach told Parsons. "I'm cutting him off."

Towsurfers retook the lineup from the solo Jet Skiers, but the scene remained distressingly chaotic. Greg and Rusty Long watched in disbelief as Jet Skis *ran over the heads* of wiped-out surfers.

The captain of the *Condor Express* took up a position barely off Larry's Bowl, a spot that had yet to show its fearsome wave, but was surely capable of doing so at any minute. When Sean Collins radioed him and said, "You're putting your passengers at great risk," he was ignored. A couple of wide sets thus nearly rolled the giant whale watcher. Then, during a longer lull, the *Electra*, a boat whose expeditionary force had been funded by Red Bull energy drinks, drifted into the impact zone. Before anyone realized it, she lay broadside in the residual white water of a broken wave.

Steve Long hailed the captain. "What the shit's going on?"

The boat had lost power.

Long wrapped a line around a cleat and threw it to one of *Electra*'s deckhands. The cleat tore right off. The panicking deckhand then threw Long a line. With a set looming, Harrington feathered his throttle while Long pulled a seventy-five-foot-long boat to safety with his bare hands.

The towsurfing teams laughed and berated one another in equal measure. They were all friends, more or less, and the peaks were perfectly shaped and utterly rippable. Garrett McNamara was one part poet and one part hellman in describing the Bank to journalist Michael Kew, who was on his boat. "The kelp looked like a woman's hair, and the submerged reef was her body," he said. "It looked very inviting. Cortes is the place to catch a 120-foot wave—not a 100-foot wave, but a 120-foot wave. I want the 120-foot wave. That way, there's no ifs, ands, or buts about it. After surfing there, I've confirmed my thoughts: Cortes Bank is the smoothest and most incredible wave I have ever surfed. The biggest wave in the world will be ridden there."

Most didn't know it, but Garrett McNamara had actually organized a loose contest for the day. It was a sort of skins game he called "If It Can't Kill You It

Ain't Extreme." Each participant—McNamara, his tow partner Ikaika Kalama, Burle, Skindog, Flea, Barron, and Loya—threw a thousand dollars into a pot, and then the boys would dissect the footage later over beers to decide who won. Thus was launched a crazed aerial assault, with surfers sometimes towing *at* the smaller waves just to see how high they could go.

Gerlach eventually towed Parsons into a barreling beauty—the wave of the day, but really, they were embarrassed. Greg and Rusty didn't even surf.

"You know what it is?" Brad Gerlach says. "You bought this house out in the country and you're like, *yeah*. I finally found some peace and quiet. Then two years down the track they're like, 'We're going to put an eight-lane freeway through your front yard.'"

Bill Sharp was more succinct. "I came to call it the Cortes Wank."

At the end of the day, on the ride back with Rob Brown, Sean Collins sat under a black sky, in a dark mood, watching as Shane Dorian powered a Jet Ski back to land in their wake. Collins thought about how easily someone could have died out there, and he made a decision. He would issue no more public alerts of Cortes Bank swells. He would do the same with Jaws. If you understood the basics of forecasting, and knew where to look, you could probably make an informed forecast by yourself, but Collins wasn't going to spray paint the data across Surfline's home page anymore. Some places were just a little too sacred and too damned dangerous to bring a circus.

After Cortes in 2004, big wave surfing seemed to hit a crossroads. Tow-surfing was *blowing up*, and every month it seemed there was a new big wave discovery. A nascent group called the Professional Towsurfing Association planned to launch a *World Tour*. Garrett McNamara foresaw a day when the sport would be as big as Nascar and surfers would boast sponsorships from Tide and Budweiser. In 2002, Dana Brown's *Step into Liquid* landed Jaws and Cortes Bank on the big screen. In 2003, Bill Sharp's *Billabong Odyssey* brought Mike Parsons's Jaws and Cortes monsters to life in sickening detail, while in 2004, Stacey Peralta and Sam George's Sundance darling *Riding Giants* chronicled the history of big waves through the lives of Greg Noll, Jeff Clark, and Laird Hamilton.

On the same swell that had spawned the Cortes Wank, forty-two-year-old Team Strapped founding father Pete Cabrinha would set a new world record at Jaws, riding a wave deemed *70 feet high*. When Cabrinha won, he hugged his wife for an eternity, held the oversize check over his head, and shouted, "I don't care what anybody says: This is a big deal. And it's a big deal to me."

And it was a big deal. Big wave surfers were garnering unprecedented coverage in the mainstream media, including the *New York Times, NPR, Vanity Fair, Outside, ESPN, 60 Minutes*, and *Dateline NBC*—to name a very few.

The only trouble was, all the attention and notoriety wasn't unalloyed good news. As the January swell made clear, one thing it resulted in was more crowded lineups, and in towsurfing, this was an exponentially more noisy and dangerous situation. How long before a novice towsurfer died, or before a novice ski driver decapitated a pro? Or, for that matter, before a pro decapitated a pro? Through the mid-2000s, the question increasingly turned to asking: Who was to blame? Was it Sean Collins and his forecasts? Bill Sharp and his big money contests? Laird Hamilton and his Team Strapped towsurfers? Was it simply the magnetic attraction of the superhuman feats of Mike Parsons, Brad Gerlach, and Greg Long?

A week after the film *Riding Giants* was released, *Time* magazine published an article that gave the first hints of a theme that Laird Hamilton would faithfully carry for the next six years. In "When the Surf's Way Up," Terry McCarthy wrote: "Others question whether the pressure of sponsorship and competitions is pushing some big-wave surfers dangerously beyond their abilities. Hamilton, who surfed Jaws reef the same day Cabrinha set the record, thinks he might have ridden some even higher waves. But he declines to enter the big-wave competitions because he thinks they are bad for the sport. 'I resent the whole concept of a bounty to try to ride an 80-ft. or a 100-ft. wave. You are provoking people that maybe shouldn't be out there.'"

This point became even more personal in the provocative 2006 story penned by Susan Casey in *Sports Illustrated* called "The Jaws Paradigm." In it, Laird Hamilton, and Casey herself, pointed the finger squarely at Bill Sharp and the XXL contests for what she described as "the goat rodeo" at Jaws during a massive swell on December 15, 2004. Casey wrote:

> No, it's not the desire to ride the biggest wave ever that Hamilton dismisses—it's the fact that the bounty will inevitably attract some surfers whose main motivation is the cash. And the cruelest irony is that Jaws, his home break, is the place where the bounty seekers are most likely to catch the winning ride. . . .
>
> People had worried that the 100-foot wave prize (now called the Billabong XXL) would lead surfers into situations that were over their heads, and the chaos at Jaws seemed to prove them right.

Bill Sharp was also quoted in the article, and he defended the way the Billabong XXL was organized—restricting participants to an approved list to avoid this exact problem—but beyond emphasizing safety, the contest, he said, can't police foolhardiness and stupidity. Sharp was quoted as saying, "It's really disappointing to . . . see how some people have decided to be a little reckless. . . . It's a sport that's just not regulated—it's about freedom, and if someone wants to be a little unwise, it's difficult to stop them."

Hamilton has continued to make his argument up to the present day, as loudly and publicly as possible, reiterating it in Susan Casey's 2010 book *The Wave* (on which Hamilton partnered to help her write): Laird and his fellow Team Strapped members blame Sharp for crowding the lineup at Jaws by making big wave surfing about the bounty. Emotions over these accusations run high, even to the point of alleged physical confrontation. According to *The Wave*, during the 2009 Waterman's Ball, surfing's annual black tie affair at the Dana Point Ritz Carlton hotel, Dave Kalama supposedly put Bill Sharp in a headlock, growling, "You represent everything I hate about big wave surfing." Both Sharp and Kalama, though, have different recollections of that night. Sharp told me, "Anyone who was there knows that's not what happened. There were no physical assaults at the Ritz Carlton. But Dave *did* walk up to me, shake my hand, and tell me he thought I was the biggest asshole in the world."

For his part, Kalama said, "I think I spotted Bill walking into the room from across the way. I was a little fired up, but not in an aggressive way. I mean, I went to shake his hand and with that I sort of pulled him into a headlock—like you would with a brother or sister. I mean, I didn't know the guy. Already I was pressing the envelope for not knowing him. And I go, 'Look man, you represent everything I hate about big wave surfing.' I didn't call him an asshole. He's got his spiked hair, and he's just pushing it so hard for Billabong and just milking the thing to death. It was me being frisky. It just wasn't a big thing."

"I just don't remember ever being in his armpit," laughs Sharp. "I must have blocked it out as a traumatic experience. But his recollection of the discussion is right on."

Even though this one incident was overblown, the underlying issues invoke strong feelings because surfers believe so much is at stake. The argument reaches into the heart of the sport, and even into each individual's motivations and basic philosophy of life: In essence, why do you surf? Are you a competitive gloryhound or a soul surfer? Are you trying to make a buck—either as a mercenary or a profiteer—or are you participating in a way of life while training to have transcendent experiences few humans will ever have? Further, who has the right to promote, and profit from, the waves? The right of locals to keep their surf spots secret forms part of surfing's DNA. Once you advertise a spot, however, what rights do you retain, and for how long?

For instance, Jaws is not a spot that you would notice from the side of the winding Hana Highway. It's well hidden. If Hamilton and Team Strapped had never identified the location of the wave where they refined towsurfing, Jaws might have remained in obscurity for who knows how long. Yet in 1996, Hamilton, Kalama, and their friends made and released the film *Wake Up Call* about towsurfing at Jaws. The same is true of the Cortes Bank: If Flame and Sean Collins had kept knowledge of it entirely within their inner circle, who would have found it? Of course, sooner or later, particularly in the age of satellites and Twitter, someone eventually would have discovered and documented these waves, but maybe not. There are numerous big waves at outside reefs in the Hawaiian Islands and off the California coast that 99 percent of surfers will never know about. Indeed, Ghost Tree, a mammoth that sits right off the Pebble Beach golf course, remained an unknown wave for several years after first being towsurfed by Peter Mel and Skindog in 2002. As such, doesn't Laird feel like he's also at least partly at fault by promoting towsurfing and advertising Jaws?

It was a question I asked him in 2008. "But there was no [financial] *incentive* for someone to go charging out there [at Jaws]," Laird said at the time, adding that *Wake Up Call* "was more of a documentation. What we were experiencing and what was going on. We felt that it should be shared. It was never our intention to try to monopolize the spot. The way that we portrayed it was always in awe, and in an inspired, respectful manner, is what I felt.

"It's just that when you start promoting it in another way, then it changes the whole thing. When you start putting up bounties and stuff. Then you're going to have guys make different decisions than you would if you're in it just to ride it. If they're just saying, 'Hey, I want to go ride some giant ones,' it seems like that mindset's a little different than if you're like, 'Hey, I'm going to win this big prize.'

"It's about your intentions," Hamilton continued. "What are your objectives? Your reasons why? Not, 'Oh I'll get chicks, or it's cool or I'll get sponsored'—all the wrong ones. The only reason should be *I love it*. I have to get on it. I have to try to ride it. . . ."

On a more pragmatic level, Hamilton added that one thing that really irked him was that—prior to the arrival of XXL-hungry crowds at Jaws—he and the Strapped crew had always been essentially a self-reliant team. With the arrival of so many upstarts, a Tragedy of the Commons scene was not only unleashed, but the Strapped crew even had to rescue surfers they didn't know—frequently people who had no business being out at Jaws. However, he did concede that one good consequence of the XXL contests was that they would come to open up heaps of new waves across the world, perhaps taking at least a little pressure off his home break.

Still, even if the XXLs are responsible for much of the overcrowding, is there really any genuine difference between Hamilton and Kalama's intentions and those of Bill Sharp? Aren't both simply trying to make a living doing something they love?

"The difference between myself and a Bill Sharp is that I go out there and I risk," Hamilton said. "He's not going out there and risking. For me, at the end of the day, I feel like you shouldn't really have much to say in the thing unless you're willing to go out and put your butt on the line.

"For me, the whole thing has to do with sending the sheep into the wolves' den. It's like, line these guys up and send them out. A lot of these guys are not qualified to go out in these conditions and surf these waves. We've said it before and we'll continue to say it. This is not a game. This is not like something you just buy a license for and go out and do. It's lifelong work."

Indeed, what motivates a big wave surfer to pursue this life? It's a question the surfers themselves often have a hard time answering. Is it possible, I ask Bill Sharp, that the lure of money is enough? Has the XXL encouraged surfers to bounty hunt for big waves they otherwise wouldn't go near?

"Let me answer the question this way," Bill says, and he queues up a video that he shot in November 2008 while at Maverick's.

On the first wave, Grant "Twiggy" Baker paddles into and sticks the drop on a four-and-a-half-story bomb. Moments later, another wave looms, and Greg Long also bare hands it. But Long doesn't make it. He free-falls a third of a football field straight down and skids across the water like a polished stone before taking the entire ocean on his head. In less than a second, he is driven thirty feet deep. The wipeout is so violent that Long's aquaman lungs are almost completely deflated while an instantaneous, crushing pressure change pops one of his eardrums like a balloon. Another wave steamrolls through the lineup. Long still fails to surface.

Shouts ring out among the Maverick's water patrol. Below the water, Long is suffocating, and the forty-eight-degree water has induced a blinding ice-cream headache, while his ruptured eardrum has given him vertigo. He swims in confused loops twenty-five feet down, circling and twitching like a dying fish. A trio of WaveRunners roar into the seething impact zone. They have seventeen seconds before the arrival of the next wave. Groggily, Long grabs his half-inch-thick ankle leash and climbs it to his surfboard. He surfaces, but his emptied lungs will only accept squeaks of air. He weakly waves to Jeff Clark and somehow manages to grab his rescue sled an instant before the next impact. Steve Long, who's been watching from the boat with Sharp, drops his head and breathes a long sigh of relief. The Godfather of Maverick's has just saved his son's life.

Sharp asks, "Would you be willing to endure *that* on the outside chance that you *might* pocket the annual pay of a McDonald's manager after taxes? And would you then paddle back out the next time it got big?"

No, money alone seems a poor incentive to become a big wave surfer. There are, in fact, too many other ways to earn a paycheck through surfing that don't involve eardrum-bursting two-wave hold-downs. At best, it seems, the promise of money could only be a partial explanation and incentive for such risk-taking behavior. Indeed, it seems to me that most of these guys—including Laird Hamilton, Dave Kalama, Mike Parsons, and Greg Long—are addicted to the life, not the livelihood.

Then again, even big wave surfers need to make a living, and so, Bill Sharp asks, is it really fair for Laird to imply that money is no motivation for himself? While Hamilton is a highly respected surfer, one who has earned the right to speak out on behalf of the sport, doesn't his accusation over money amount to a double standard? Sharp wonders: Why are he and the XXL surfers accused of being sellouts when Hamilton boasts sponsors from Oxbow to American Express to Davidoff cologne and has his own line of stand-up paddle surfboards? "How much was he paid to 'pimp' Jaws in the film *Die Another Day*," Sharp asks. "Why did he split a million-dollar advance with Susan Casey to be a subject in her book *The Wave*. I don't think any other surfers she interviewed were offered a *bounty*.

"Asserting that the guys who surf in the XXL are doing it for bounty—or that I'm somehow leading lambs to slaughter—those are quotes from someone who either has their own agenda to push or does not have any grasp of what goes on in the mind of the big wave surfer. I mean, whether you're Mike Parsons, Greg Long, or even Laird, you do this because you love it. If we shut down the XXL today, on the next big swell, I can guarantee you that there are going to be no fewer surfers in the water."

I interviewed Dave Kalama about this, and he was clearly conflicted. "We made surf videos," he says of Team Strapped and he and Hamilton's own role in the chaos at Jaws. "We organized Sunny [Miller] and the helicopters to get ourselves coverage. That coverage is gonna make it crowded. That's how being a professional surfer works. No way we can say, 'No, you're not allowed,' and then we still make money off it. That would be completely hypocritical. But there's a big difference between that and standing on a podium and saying, 'Come one, come all.' Throw yourselves into the gladiator pit. We definitely showcased it, but there was no money on the line.

"When a large multimillion-dollar company says, you know what, we're going to milk this thing for every dollar we possibly can, it changes it from an individual to a corporate thing. . . . You take a guy like Mike Parsons—he's fulfilling his sponsor obligations and being able to ride these big waves—*that* I'm willing to accept, and I think most guys are. It's the guys sitting in the corporate offices and golfing, and they've got *nothing* to do with it, and they're making the most money off it. That's what rubs me the wrong way."

In fact, when he considers the surfers themselves, Kalama's tone softens. "Enough time has gone by now that I can look at guys like Snips and Gerr, and I mean, I can appreciate them rather than look at them as competition. I can look at it like, we definitely started a path and opened a door."

Then, when Kalama discusses the current crop of new-school paddlers—guys like Greg Long, Mark Healey, Twiggy, and a Maui surfer named Ian Walsh—his tone switches to reverence. "Those guys are doing *exactly* what they're supposed to do. Taking the subject matter and rolling through it with a steamroller. I'm almost glad I'm not having to do what they do."

For his part, Bill Sharp concedes that Billabong puts up the check because they recognize that sponsoring a major event like the XXL is a profitable business decision. But in the last decade or so—and it is something that Hamilton and Kalama help spark—this corporate interest has allowed at least a few surfers to actually make a living chasing adventure, satisfying their addictions, and blowing the minds of mere mortals.

"The core of all of this—the core of my life—has always been about having amazing adventures in places no one has explored before," Sharp says. "There was no such thing as towsurfing or the XXL when we first went to Cortes Bank. It was done purely for the same reason George Mallory sought to climb Everest or Laird first towed Log Cabins. We did it *because it was there*. We weren't thinking, 'Wow, here's a wave where people can win the XXL.' That's beyond absurd. The XXL is a document of what's going on in big wave surfing today. It's a mirror on the sport, not some engine propelling it in the wrong direction. And anyone who alleges that any of the top surfers are riding big waves purely for the bounty is ignorant of what goes on in the hearts of these men—and women."

Like Kalama, Hamilton's complaints about the mercenary impulse seem to be in part frustrations over the role of commercialism in surfing in general. Many surfers complain about corporate involvement in surfing, even when sponsorship is the Faustian bargain that allows professional surfers to afford to do what they do. Yet not every surfer opts in, and the living that even the most successful XXL surfers make would be considered marginal. In 2008, I was out at Maverick's aboard Rob Brown's boat with Greg and Rusty Long, and I noticed that Greg had blackened out the O'Neill logo on his garish blue wetsuit. One of the best big wave surfers on Earth, maybe *the best*, had been surfing for some time with no sponsor *at all*. Meanwhile, Greg was literally living in his beat up Ford Econoline van down by the San Mateo River, and Rusty was stretching a thousand dollars a month to pay for food, travel, rent, catastrophic health insurance, and gas for his rust-eaten

Land Cruiser. Brad Gerlach plays music and works as a surfing instructor. Skindog hustles clothes as a representative for Volcom; the most he has ever earned from a sponsorship was thirty-five thousand dollars a year. Pete Mel puts in grueling road time for main sponsor Quiksilver, as an announcer on the ASP World Tour, and helps run the family surf shop. Mike Parsons has just left Billabong for a full- time job as manager and coach for up-and-coming pro surfer Kolohe Andino. To some, it may sound impressive that Evan Slater was once the editor of *Surfer* magazine, but as any *Surfer* editorial staffer will tell you, the glory of the position is accompanied by poverty wages.

If this is the income and living that the best big wave surfers can command, how much of an incentive to others can it be? Those athletes for whom money is most important choose to compete in the small wave-friendly World Tour, or they choose another sport. Unless you're Laird Hamilton, big wave surfing just isn't a way to get rich.

Nor are Laird's and Bill's the only valid opinions on the matter. Most of the XXL surfers, not surprisingly, chafe at a suggestion that they're bounty hunters. But it's the ever-outspoken Sam George who gets at a truth that might, and perhaps will, make both Laird and Bill spitting mad.

"Not everyone can be Laird Hamilton," says George. "But I've seen Bill ride giant waves. He doesn't have to prove his physical courage to anyone. You know, it's funny. Bill would come to single-handedly create a new genre of professional big wave riders. It's something Laird Hamilton might have done, but he didn't. I'm not saying Laird couldn't have, but he didn't. And I'm not denigrating Laird or his contribution to big wave surfing. He has *completely* changed the sport. But the way Bill would come to revamp big wave riding—and put the focus back on it. He's almost solely responsible for the fact that Laird Hamilton has a career in big wave surfing. Without Bill Sharp turning the world's attention back to big wave surfing, Laird Hamilton would still be out riding Jaws. You just wouldn't have seen him doing it on the cover of *National Geographic*."

What Sharp really did to big wave surfing, George says, and what some self-described purists consider an almost unforgivable offense, was to do away with Buzzy Trent's enduring romantic notion that "big waves are not measured in feet, but in increments of fear."

George says, "Bill cut away all the bullshit and upped the ante. He said that if you rode it, it should be measured, and you deserved to be recognized. So if you really think about it, those two in the end, they're both responsible for all this. Laird and Bill. They're a team. It's true. If not for Bill, there's no Laird. If not for Laird, there's no Bill."

Chapter 11:

TRIFLING
WITH THE
ALMIGHTY

"What is it, what nameless, inscrutable, unearthly thing
is it; what cozening, hidden lord and master, and cruel,
remorseless emperor commands me; that against all natural
lovings and longings, I so keep pushing, and crowding, and
jamming myself on all the time; recklessly making
me ready to do what in my own proper, natural heart, I
durst not so much as dare?"

—Ahab, from Herman Melville's *Moby-Dick*, 1851

Following his sobering wake-up call at the Cortes Bank in January 2004, Sean
Collins remained true to his word. He continued to include the Bank in his wave
models, but he, and everyone who went out there, basically quit talking about it.
The change was stunning.

It wasn't like Collins was the only one who knew how to stitch chart data
together. Anyone could have made reasonable predictions of the winds and conditions with a little effort. It's thus downright amazing to consider that between
2004 and 2008, only eight to ten surfers ventured out to the Cortes Bank. Perhaps
seventy-five people and athletes showed up in a single day in January 2004, and
perhaps two million surfers live within a hundred miles of the Cortes Bank. It
wasn't like big wave surfing suddenly became unpopular, or the XXL contests were
canceled. Billabong's half-million-dollar prize for the first person to ride a 100-foot
wave still stood, awaiting a taker. The sudden lack of surfers at Cortes, then, would
seem to underscore a few simple facts: Without a heaping dose of spoon-fed data,
an ability to quickly leap through a small open window of opportunity, a captain
who knows the waters, and cojones the size of boulders, the Bank is just one of the
most dangerous, difficult places on Earth to ride a wave.

The surfers who *did* venture out were, and largely are, part of Sean Collins's
inner circle: Gerlach and Parsons, the Long brothers, and a small smattering of
Hawaiians. Collins continued to make less-obvious forecasts for Jaws, and the

break eventually became somewhat less crowded. Even without surf alerts, Jaws remained a whole lot easier to access, while the Bank became a sort of private terrordome.

In fall 2004, nine or so months after January's Cortes Wank, the first secret Bank mission was launched, and it included Steve Long, Bob Harrington, Greg and Rusty Long, Mike Parsons, Brad Gerlach, and Rob Brown. The swell was enormous and the weather perfect as their two boats rounded Catalina. But just off San Clemente, the teams disappeared into a thousand-mile-wide bowl of pea soup fog. The journey became a chilling trip into inner space. Steve Long and Harrington tried to give the entire Bank a wide berth, but huge swells do strange things when they interact with a fourteen-mile-long mountain. All around Bishop Rock, there was foam from broken waves scattered across the sea surface, and deep thunderclaps from unseen breakers. Steve had nightmarish visions of rogues rampaging over the boat from deep in the mist. "It was just giant," he says. "We were scared shitless. We got separated from Rob and his boat and never found them again."

Steve Long and Bob Harrington turned back.

Meanwhile, Rob Brown's hunting party—Mike, Brad, Rusty, and Greg— motored in close to Larry's Bowl. The waves were so big and their detonations churned so much air that they created a small circle of clear weather above the impact zone—something like the effect of a cold downdraft at the base of a foggy glacier. Once on a wave, you'd be able to see, but it was too foggy to track the sets as they approached or estimate how big they really were. You might be sitting way out there, with *Phantom of the Opera* pipe organs playing in the background, when a 60- or 70-footer stormed in before you even had a chance to react. Caught inside, you'd be pushed into the foam and then disappear into the fog—where you might simply never be found.

As Bob Parsons might have said, Mike and his friends were brave, but they weren't foolish. No one left the boat.

In late February 2005, a tightly wound fifty- to sixty-knot storm swept across the Pacific. A small cluster of larger boats reached Bishop Rock carrying Rob Brown, Sean Collins, Mike Parsons, Brad Gerlach, Greg and Rusty Long, Brazilian fiend Carlos Burle, and a young Hawaiian named Jamie Sterling. Cortes was in a foul mood. A stiff, frigid breeze blew from the northwest, turning the ocean gloomy, lumpy, and mean.

Bill Sharp was also there, having hitched a ride with an audacious good ol' boy surf photographer named Les Walker aboard a tiny twenty-one-foot bass boat that Walker had christened *Fried Neck Bones*. The boat was a dinghy compared to

Rob Brown's bomb chaser and far, far too small to safely make such a journey, but she held ample gasoline and beer. In addition, *Surfer* magazine's newly minted editor, and Parsons's boyhood chum Chris Mauro, had begged for, and was given, a spot on board.

The first twenty-second waves staggered across Bishop Rock at around 10 A.M. The winds briefly laid down long enough to lure everyone out, then they turned back on, offering a 60-foot-high mogul field of nightmares. "It was the fucking real deal," Sharp says.

Everyone was towsurfing. Gerlach and Parsons traded big, open-faced turns while Greg pulled Rusty onto a horrendous barreling thing that sucked Rusty over the falls and drove him down painfully deep. Rusty called it quits.

A crew of documentary filmmakers hired by Red Bull convinced Jamie Sterling to don a helmet cam. (The forecasts might have been private, but the results of Cortes missions were still invaluable editorial property.) The goofy-foot struggled to hold a terrifically difficult backside line through the chop. His reward was a forty-five-knot face-plant and a trip through the meat grinder. The camera was *gone*.

Parsons idled over on a Jet Ski and ordered Mauro to take the tow rope. Chris was terror-stricken. He'd only come to watch and report for the magazine. He'd never seen Cortes Bank or even towsurfed before.

Parsons wasn't *trying* to terrify Mauro. He simply, sincerely wanted his old friend to experience the grand magic up close. That didn't stop Mauro from experiencing a frightful déjà vu he'd not felt since Mike ordered him out into huge, predawn Sunset Beach as a kid. After Mauro strapped in, Parsons first tried to set him up for a tough, slingshot whip from behind the ski. Mauro instead dropped the rope and saluted his old mentor with two raised middle fingers. With feet still strapped to the board, he slowly sank to his chest. "I'm out here alone at Cortes, flipping off Mike, and sinking in the middle of a fricking set," he says.

Snips gave Mauro just enough time to stare death in the eye before hauling him out and tracking down a somewhat smaller wave. He would ease Mauro in from behind onto what was still the biggest wave of his life. When "Sis" let go of the rope, he thought he was fine—for a moment—then the wave stacked to vertical on Larry's Bowl and launched Mauro into space. He hit the water like a slab of steak chucked onto a sidewalk before being buried alive. Chris was no longer a full-time surfer. He was a desk jockey. An editor. He had a little baby. As he tumbled and rolled, he thought, *What the hell am I doing?*

Mauro laughs. "After twenty years, I had just paid for all Mike's pent up frustrations with me."

Sharp, Walker, and Mauro motored back that night in a light fog. They watched in awe as small spinner dolphins played off their bow wake—

bioluminescent organisms turning them into sparkling purple torpedoes. "Then we just went into this area that was an electromagnetic freak show," says Sharp. "The compass started spinning and all the electronics just went zonko."

Walker piloted dead ahead into the emptiness and promptly freaked out several minutes later when *Fried Neck Bone*s crossed the lingering phosphorescence of her previous bow wake. They were in a Pacific Ocean Bermuda triangle. "Then a few minutes later we just sailed out of it," Sharp says. "You always hear stories about places with these magnetic anomalies, but you never actually expect to see one. It was terrifying. Like something out of *Poltergeist*."

By fall 2005, it was clear that Larry Moore was losing his battle. Right from the moment of his December 2002 diagnosis, Flame fought his brain tumor with the same relentlessly sunny intensity he put into battling fellow photographers. He egged his doctors on, demanding more radiation, deeper doses of chemo, and risky surgery. After better than two years of this hell, he was still choosing photos—now alongside editor Evan Slater—from his old perch at *Surfing*. One day when George Hulse stopped in, Flame excitedly described the incredibly precise and noninvasive "gamma knife" that would be used for his next brain surgery.

Flame kept working longer than he probably should have. He started to fall frequently and wrecked his car more times than his wife, Candace, could count. Eventually, a reaction to an overdose of chemo nearly killed him. When he came home for good, Hulse became a regular confidant. "He was reaching out for companionship and friendship, and of course our Christian faith was a bond," says Hulse. "But I mean, he had been my mentor. Had such an effect on my life. I was just humbled and honored that he reached out for me."

They prayed and laughed a lot about the past. The memories of surfing, of course, took center stage. Hulse had first been introduced to Flame by Brad Gerlach on a perfect afternoon at Trestles. "I got a sequence in the magazine," Hulse says. "And Brad just heckled me, like, 'Aw man, how come I didn't get the shot?' Then there was a time, just Flame and I went down to this beach break in Mexico. I hooked up into a barrel—and he was right there. He goes, 'I think that's a cover shot.' And it was."

Sometime after their first trip to the Cortes Bank, Hulse made a journey out to a secret wave on the western edge of Kinkipar with Flame, Chuy Reyna, Dave Parmenter, and Gary Clisby (owner of the *Pacific Quest*). Everyone had been surfing well outside when Hulse noticed Flame lying on his surf mat with his head down. He paddled in and asked how Flame was. "He goes, 'I'm seasick. I'm puking.' I said, 'You going in?' 'No, I came out here to get shots.' So I decided to surf the inside bowl with him. I did this kind of low hook with the lip of a wave coming over me. When I made it back out, he goes, 'You won't believe the shot I got while I was puking.' That was my last *Surfing* cover."

Flame's final days were, of course, hell on Candace and Colin, and Flame knew it. He had a goodly number of friends—grown-up Salt Creek rats—to help out where they could around the house, but they could only do so much. In the end, Candace's dad reeled in much of the slack. Her father had once been a stern taskmaster, not open with his words. "But he became Larry's appendage." Candace says. "And the experience—it just changed him—changed him entirely. I mean, that's what Larry did—he changed people's lives."

"I want you praying for my wife," Flame told Hulse at the end. "And I want you to speak at my memorial. I want you to tell all the people there. Tell them that I know I'm going home to be with the Lord and that I'll be okay. I want you to get up and tell them that."

Larry "Flame" Moore died peacefully in his home on October 10, 2005, and George Hulse fulfilled his promise.

Six days after Flame's passing, a dawn paddle out in his honor was planned at Salt Creek. The night before, lightning and thunder shook the foothills of southern Orange County, and a notice was sent out that the event would have to be postponed. Bill Sharp curses the fact that he was among those who heeded it. Yet George Hulse, Mike Parsons, Evan Slater, and a great many others didn't get the message. As maybe 150 surfers hit the water, a hole appeared in the clouds, bathing Salt Creek in perfectly front-lit "Larry light" and enshrining everyone beneath a rainbow. "There was definitely some higher power at work," says Slater.

The following New Year's Day, Candace carried Flame's ashes down to Salt Creek to cast them into his sacred waters. As she walked along the sand with Flame's sister, Celeste, Candace noticed that they were sort of being followed by a black balloon. They walked out onto the rock jetty, which had served as a base for thousands of Flame's photos, and released his ashes. The balloon was eventually blown out alongside them into the cold ocean. They watched it bob through the waves until it was just outside the breakers, and then it just took off and flew away in the direction of the Cortes Bank. "And we just watched it and watched it," says Candace. "Until it was out of sight."

A single great storm would follow in the wake of Flame's death. It blasted across California, leaving but a small window of big, clean surf down south. On December 20, 2005, Greg and Rusty, Snips and Gerlach bounced down to Todos Santos with Rob Brown. It was shockingly big, and everyone paddled out but

Gerlach, who was not too proud to admit that the very idea of paddle surfing that swell scared the shit out of him. "It was like being at a strip club," Brad said. "Look, but don't touch."

Gerlach's trepidation was validated when a 50-foot set cleared the lineup. Skis were thus fired, and Todos grew and grew, until it was performing the best impression of Cortes Bank anyone had ever seen (though Sean Collins estimated that the waves at Cortes would have been around 30 percent bigger). At around 3 P.M., Parsons towed Gerlach onto a wave that made him look like an ant on a halfpipe. The following spring, Gerlach would take home his first Wave of the Year XXL Award and a check for $68,000. He missed breaking Pete Cabrinha's world record by two feet.

A couple of months after Brad's ride, in early 2006, very long-period ripples borne of a distant Alaska-bound low again pulsed toward Cortes Bank. Parsons, Gerlach, and Brown found empty waves perhaps 15 to 20 feet on the face. They were all alone. "I was just putting my board on a rail at a thousand miles an hour," says Gerlach. "It was big, it was slopey, sometimes it was hollow on Larry's Bowl. Dream surfing. Just us, out by ourselves at our favorite spot in the world."

They said a prayer for Flame, hoisted beers in his honor, and motored home.

Perhaps in deference to Flame, the swell window at Cortes would not reopen for almost two more years. The 2006–07 season passed without incident at the Bank, but the winter of 2007–08 opened promisingly. By late November, Hawaii had been pumping and Maverick's had already awoken three times. Sean Collins attributed this to an intensifying cold La Niña episode. While the literal polar opposite of a warm El Niño, extreme sea-level temperature contrasts found during a La Niña can still trigger massive, violent wobbles in the jet stream. The same could be summarily said for big wave surfing that winter, which became a watershed season that careened between unprecedented highs and sobering lows.

During the last week of November, the season's first truly great frigid blast tumbled off Siberia and was pumped up with the warm steroids of a dying cyclone in the lower latitudes. The storm wouldn't actually begin to peak until it was due north of Hawaii. Thus, when Maui was blitzed on December 3, the swell angle was too north for Jaws. Laird Hamilton and his friend Brett Lickle instead made a stand a few miles off the coast at a pyramid-shaped wave they call "Egypt." After Lickle hurled Laird onto a wave that he told me was surely better than ten stories tall and the biggest thing he had ever surfed, the duo was overtaken by another wave, despite motoring flat out at fifty-five miles an hour. They were obliterated. The aluminum fin on Hamilton's free-flying, twenty-pound board then flayed Lickle open from his Achilles tendon to the back of his knee. Despite surfacing amid three vertical feet of choking foam that had been painted pink with Lickle's blood, Hamilton coolly and heroically ripped off his wetsuit, then fashioned a tourniquet, swam a lonely, naked half mile for the Jet Ski, and raced back to shore ferrying Lickle before the life completely bled out of him. It was not only the heaviest session of their lives, but also a validation of the core gospel Laird had been preaching from the temple of Jaws for a decade: Big wave surfing is not a game or a contest or a lark. And it doesn't matter what technology you have or how skilled you are, any wave could be your last.

The mammoth storm continued to develop unusual characteristics—particularly an incredibly broad wind spread that led computer swell models to paint the eastern half of the North Pacific in a great violet orb. "We had a complicated forecast for that storm," Scripps researcher Bill O'Reilly told me a week after the storm. "You'd usually see something like that in an El Niño year, but we're seeing a strong La Niña. I think some are still scratching their heads over how and why it developed like it did."

A couple of days later, the storm's swells destroyed the Southeast Papa buoy eight hundred miles northwest of San Francisco. Before Papa blinked out, he was rolling above seas with significant heights of better than 50 feet. This means that the highest one-third of the swells were around 50 feet, while a great many others, perhaps ten percent, were far larger—in the 80-foot range. According to Sean Collins, the Papa buoy was very occasionally, but most assuredly being buried by waves better than 100 feet high.

"You go out in the midst of a storm like that, and you happen to be in the wrong place at the wrong time—those waves get in phase and you find yourself in front of the highest one-tenths—at that point you're in a world of hurt," said Collins. "People just aren't usually there to see it. When they do, it's a huge phenomenon—that's when you have a situation like the *Andrea Gail*. But really it's a common occurrence. It happens all the time."

A giant swell like this puts Collins in a touchy situation. The first time I ever fully appreciated this strange, central dilemma of his life also happened to be the first time I ever saw a giant wave in person. I traveled out to Todos Santos in February 1999 with Collins and Surfline's Dave Gilovich in a primitive attempt to webcast the Reef@Todos big wave contest for Surfermag.com.

Collins had been responsible for the forecasting call, but as our boat rounded the corner, we saw nothing but a few surfers bobbing off a nasty crag of rock in a serene, open ocean. Tens of thousands had been spent by sponsors and poverty-stricken competing surfers, and it was flat as a lake. The God of Surfline cursed himself, but kept anxiously repeating, "It's gonna show. It's *gotta* show."

Suddenly—and I'll never forget this as long as I live—a humpbacked shape appeared off the rocks. A dreadlocked South African named Cass Collier dug hard, and this *wave*, the scariest damned thing I had *ever* seen from land or sea, swept beneath him at warp speed. I had no inkling a wave could move that fast. And I had no idea a human was capable of negotiating such a roaring, beastly thing. Seeing a 30- or 40-foot wave in a photo or video is one thing. Seeing one up close, watching that spray, and hearing that primal roar is just an other-worldy experience. Cass Collier stuck his drop. Collins breathed a sigh of deep-est relief and looked like he'd just ridden his own 20-footer. The contest was on.

Fast forward a decade. As a titanic swell sweeps the Pacific, Sean's phone lights up with calls from guys like Greg Long, Mike Parsons, and Laird Hamilton, all of whom want good, big, empty waves and need his advice. Sean doesn't want to send hordes to Jaws, Todos, or Cortes Bank, nor does he want to throw a good friend off the scent. Like any newsroom editor, Collins knows the value of a Surfline exclusive, but on the other hand, Sean has a business to run broadcast-ing surf forecasts to the world. If he plays up a swell too much and lineups get crowded, he might catch hell—from Laird in particular. If he misses, or worse, undercalls a forecast, Surfline subscribers raise hell because he's "blown a call."

The purists who hate Surfline would say that's just as well. They blame Collins for the crowds, anyway. And indeed, I don't pity Collins, nor does he ask for sym-pathy. Being surfing's fiber-optic kahuna is a pretty decent life—something surf-ing's "beach bum" forefathers could have never imagined. Consider this, however: Sean's mania for forecasting is the result of the same gnawing inquisitiveness that led him to sit on his roof endlessly counting and timing waves with a stopwatch, to carrying a huge old weatherfax machine into the Mexican desert and juicing it with a car battery, and to the later founding of Surfline. These were pioneer-ing efforts, and they arise from an impulse Sean Collins can't control. Some call Sean an Ahab, but really he is no less a high-sensation seeker than any other big wave surfer he has ever spilled the beans to or argued with. He finds the same sort of rush in nailing the forecast for a great Pacific storm as Laird Hamilton, Mike Parsons, and Greg Long do in reaping the benefits of his knowledge. Plus, Sean Collins probably rides more empty waves than you, I, or anyone we know.

On December 4, Greg Long, Mark Healey, and Pete Mel awoke to find Maverick's enshrouded in a death veil of fog. Long and Healey had surfed giant Waimea Bay the day before and were still jet-lagged from their flight. They had been receiving regular updates from Mike Parsons, who that morning had tar-geted a recently discovered towsurfing spot in Carmel. The wave was another freak of nature that on precisely angled west swells shook the earth in front of a beach strewn with VW Beetle-size boulders. Amazingly, it broke within a nine-iron shot of the Pebble Beach Golf Course. The spot's proper name was Pescadero Point, but a Carmel local and Maverick's veteran named Don Curry named it Ghost Tree, after the bleached, gnarled skeletons of Tecate cypress that loom above the break.

Skindog and Peter Mel had been first to tow Ghost Tree back in 2002. For better than two years, everyone had managed to keep the wave under wraps—a feat on par with hiding King Kong behind a line of ornamental shrubbery. The wave promised to be every inch the equal of Maverick's, and its discovery augured well for future finds along the California coast. On December 4, 2008, Parsons sent word to Long that conditions off Pescadero Point were sunny, perfect, unbelievable—and crowded as hell.

Long and Healey bombed south to catch up with Parsons at Ghost Tree, while Pete Mel and his friend Ryan Augustein tentatively motored out in the fog to Maverick's. It was as mean as they had ever seen. Grant "Twiggy" Baker and Grant Washburn were also towsurfing together at Maverick's, and they were alternating between amped up bliss and mortal terror. At one point, everyone watched in horror as Darryl "Flea" Virostko was launched by a wave that seemed to have been forged by Neptune himself. Yet everyone was astonished when Flea emerged from his wipeout not only conscious but ready to go again. He would later state with certainty that someone had grabbed him and brought him to the surface. Mel was clean and sober by this point. Flea was not. Had he not been sizzling on meth, the outcome of the wipeout might have been different. Of course, had he been thinking straight, he might not have gone in the first place. "I basically ate shit," Flea would tell a stunned anchor Julie Chen the next day on CBS's nationwide *Early Show* broadcast. Augustein soon dragged Mel onto an equally deadly wall that offered a hellish beat down. "I thought it was going to rip my limbs off," Mel says.

At around 12:30 P.M., a rescue boat motored by in the fog, asking Mel and the other surfers if they'd seen a vessel in distress. A crab boat called *Good Guys* had gone down, and her two crewmen were lost to the waves.

Meanwhile, at Ghost Tree, everyone was towsurfing except for two surfers, a Monterey local named Peter Davi and a buddy of his named Anthony "Tazzy" Tashnick. Davi was one of the only surfers to ever dare paddle at Ghost Tree, and he generally took a dim view of towsurfing. He and Tashnick were straddling traditional big wave guns among the watercraft-bound mob, and Davi was not giving an inch. He was big and rugged—a Laird Hamilton–size commercial fisherman whose deep Italian roots extended from Sicily to the North Shore of Oahu to the old wharves of Cannery Row.

Pete Mel called Pete Davi the kind of guy you didn't want to cross, but if you became his friend, his generosity was literally without limit. He'd find so much jade along the Big Sur coastline that he was just constantly giving it away—even to people he barely knew. Davi had been something of a mentor to Mel, Flea, and many of the other Santa Cruz boys, paving their way onto the North Shore of Oahu. "We called him Pipeline Pete," Mel says. "He brought me in and introduced me to all the North Shore guys you feared most in my early days out there. That was just an incredible benefit to me."

Both Parsons and Gerlach knew Davi from the North Shore. Mike had always been particularly impressed with his charges through seemingly impossible Pipeline barrels on a bright green board. "Pete got big ones," Mike said. "He charged. *Charged.*"

The morning Mike Parsons and Brad Gerlach were warily towsurfing Ghost Tree, the waves were frightful and as crowded as the worst days at Jaws or the Cortes Bank. This was due in equal parts to the wave's stupendous barrels and the fact that NOAA was finally about to make good on a no Jet Ski policy that would extend along the California coast from Maverick's through Carmel and down beyond Big Sur. This might be everyone's last chance to towsurf. On their very first wave, Mike Parsons slung Gerlach onto a small "shitty" warm-up wave and Brad kicked out right next to Pete Davi. "He was just right there laughing at me," Gerlach says. "I kind of laughed, too. Pete was a pretty cynical guy. I took it like, 'Wow, that wave was a piece of shit. Why are you towing around catching a piece of shit like that?'"

After trying to scratch into a few himself, Davi realized that the only way to ride Ghost Tree today was towsurfing. He seemed amped up and agitated. If Pete Mel had been there, he would have instantly recognized why. Davi told his buddies Anthony Ruffo, Kelly Sorensen, and Randy Reyes, "I'm forty-five years old, and I want one of those fucking waves."

Reyes towed Davi straight off a ski he was sitting atop and out onto what would be his first and only wave. It was no monster but still plenty tough to ride on an 8-foot 6-inch paddle surfboard. Ruffo was sure Davi would be pinched between white water and rocks, but Davi knew Pescadero Point like the back of his hand and rocketed safely around the corner. "Everything was cool," Ruffo said. "We laughed about it."

After the wave, Davi didn't return to the lineup. Despite offers for a lift back to shore, he instead turned to make the nearly mile-long paddle back to Carmel's Stillwater Cove, a deep, wide embayment of vicious currents and abrupt shelf rocks. He tried to catch a smallish wave into the cove by riding around the inside edge of Pescadero Point, but he went down and his ankle leash snapped, sending him onto the rocks. Without his board, he was left to swim alone toward the beach. After he had made it fairly far into the cove, a local saw him rise on a swell that was about to crash down onto a slab of reef.

An hour later, the news rippled through the lineup: Anthony Ruffo had stumbled on Davi's body. A posse of Pete's buddies formed a circle in deeper water off the wave's edge and lowered their heads in Pete's honor. Some surfers ignored the news altogether and went on blithely doing their damnedest to kill themselves—an act Davi's friends considered a pinnacle of disrespect. After hearing the news, Parsons and Gerlach motored over into the deep water, badly shaken, and talked with several of their friends. Parsons in particular felt a painful déjà vu. Should they get out? Quit surfing? What was the best way to honor a guy they knew as a maniacal charger?

From the outside, it may be hard to fathom the reaction of the surfers that day, particularly those who didn't even pause. Most of the other surfers, even after sending Davi their prayers, returned to surfing. Yes, the swell was a rare monster, but had big wave surfing reached such a manic, fever pitch that even death was skipped over like a boil at the bottom of a wave?

"We really talked about it," Parsons said. "It was one of those situations where, we decided, if it happened to me, I want my friends to keep surfing. I mean, if it's closing out at Waimea, and you decide to get out—where it's just way too heavy for everyone—that's one thing, but that wasn't the case. It was huge, unbelievable, perfect. I know some people bailed who were really close to him, but everyone dealt with it in their own way. Most of the guys felt that the best way to honor Pete was to ride an even *bigger wave*. That's what Brad and I decided—let's ride one for Pete. I mean, when Andy Irons died in 2010, Greg, Shane Dorian, Mark Healey— we were all Andy's close friends, but we still went to Cortes Bank. Andy wouldn't want us to miss an insane day. Neither would Pete. They'd want us to pull in."

By the time Skindog, Greg Long, and Mark Healey reached Ghost Tree, the mood among the remaining surfers was explosive and somber. Parsons eventually honored Davi by towing Gerlach into a hellbender with a barrel almost comical in scale, but every other wave, even the most out-of-control rogue peak, was spoken for. When Healey bunnyhopped an exposed boulder on a wave Greg Long hurled him onto at around fifty miles an hour, the cushion of water thrown out by the explosion was all that saved him. After that, Gerlach, Parsons, Healey, and Long packed it in and headed south toward Todos Santos.

When word of Davi's death reached the parking lot at Maverick's, Peter Mel, Flea, Garrett McNamara—all the guys up there—were extremely upset by the death of their friend. Flea was particularly shocked, since he'd only that day survived a wave he was certain had killed him. He would tell Pete Davi's son Jake that he was not only certain that someone grabbed him and pulled him off the bottom, but that someone had been Jake's dad.

A month later, Mel, Flea, and Peter Davi's friends and family were doubly crushed when a toxicology report showed what many had quietly feared: Davi had methamphetamine in his bloodstream. Had it led him to make a decision that would kill him? No one will ever know. If any good could come out of such a tragic loss, perhaps it was the eventual redemption that would lead Flea to his own crusade to help others kick the stuff.

"That damned meth," Mel says. "It's just so addictive. So fucking gnarly. I mean, it caught a few of us just so off guard. When I look back, I realize how lucky I was to pull out. Because it almost destroyed everything in my life, too."

The next morning, Greg and Rusty Long, Mark Healey, Mike Parsons, Brad Gerlach, and photographer Jason Murray reached Ensenada on their way to Todos Santos. Winds offshore would be too heavy for a session at Cortes Bank, but as the swell ran southward down the Pacific Coast, Todos Santos should be just as epic as Maverick's and Ghost Tree. But by the time they reached the lineup at Todos, the exhausted crew was running on adrenaline fumes. Offshore, they found Evan Slater and Colin Smith, who had motored down from San Diego to paddle surf, and a few towsurfers already in the water. When Healey, Slater, Parsons, and the Longs paddled out, the verdict had been rendered: No goddamned towsurfing till the paddlers were through. "You can't be towing through a committed crew," said Rusty. "That's the cardinal rule."

Gerlach was happy to sit and watch, but though Mike paddle surfed, he wasn't feeling it. "I was like, if I survive this, I'm good," Parsons says. "But Greg, Rusty, and Mark—they were just in a zone."

Greg paddled well outside. As if on cue, he managed to strong-arm onto a boiling slab that Jason Murray called 45 feet high. Then at around 11:30 A.M., a set loomed the likes of which no one, not even Parsons, had ever seen from the deck of a paddle surfboard. Snips and Greg desperately clawed to the top of the first wave, trying to get over it, and looked back over their shoulders. They were a hair's width away from being flea-flicked six stories straight down. "I was done after that," Mike says. "*Done.*"

Parsons left the lineup, and then, Greg says, "It just got bigger and bigger. The biggest I'd seen in my life. We were *paddling* and I swear it was bigger than the tow wave Brad won the XXL on the year before."

Brad agreed. This Todos swell was even bigger than the one that had produced his 68-foot XXL winner in December 2005.

Greg somehow dug into and rode a beast that was later determined to be 53 feet high. Afterward, as he returned to the lineup, another giant set loomed. Everyone went into a dead sprint, again paddling over, not into, the waves. "I was terrified," Greg says.

Rusty actually tried to surf the second wave in the set, but he didn't quite catch it. At approximately 65 feet, it would have been the greatest wave anyone had ever paddled into, but Jason Murray is quite sure that Rusty would have died. So is Rusty. "It would have been hideous," he says. "I had a little guiding hand pulling me back."

Unfortunately, as Greg feared, not everyone made it over, and Healey and Evan Slater were among a group that took a watery skyscraper directly on the head. Murray roared in on a Jet Ski to find a shipwreck zone of broken boards and half-drowned bodies.

"Everybody was sitting out the back, after that, really spooked," says Greg, musing on this three-day run between Waimea, Ghost Tree, and Todos. "We packed it up. I went in, had a margarita, and thought about the wildest three-day swell chase of my life."

The session was a bellwether—arguably the biggest paddle session in history—and a few things were becoming evident to Greg, Twiggy, Rusty, Healey, and the core group of hellmen who were their friends. "The idea kind of came to us—that there was something kind of *missing*," says Greg. "That we really hadn't explored paddle surfing as thoroughly as we could."

After all, dedicating yourself to towsurfing was damned expensive, and it just didn't get really, truly *towsurfing* big terribly often. If you trained yourself to hold your breath for four, five minutes, until you found the point where you blacked out; if you practiced hard-core yoga to prepare for twistoflex hold-downs; if you had a couple of towsurfing skis to run rescue—then perhaps you could both ride and survive waves now considered only even *possible* from behind a Jet Ski. It was, as they say, a *wake-up call*.

And the session really got Greg in particular to wondering: Had Evan Slater and John Walla been dinosaurs or were they actually ahead of their time? Could you paddle in at the Cortes Bank when it got big? A few weeks later, a Pacific storm forecast seemed to promise the chance he hoped for. Yet it soon became evident that maybe he shouldn't mothball his tow surfboards just yet. Looming well out over the horizon was the greatest storm Greg Long, Mike Parsons, and indeed the whole West Coast had ever seen.

To this point, the 2007–08 winter season had produced a historic run, even if it had been marred by a death and several near-tragedies. Yet for some it wasn't enough.

"I'm fiending—completely fiending to go out to Cortes," Mike Parsons told me one afternoon that December. "I'm constantly looking at the sun and the conditions. I'm obsessed with it. When is it going to happen? I think Greg's probably the most like me. We're just wondering, why can't the swell be *today*? But then you have to remind yourself, that's exactly why it's the coolest place in the world."

Every day after the mist settled following that early December storm, Mike Parsons and Greg Long checked the weather models with an eye on Cortes. Then, a few days after Christmas, Sean Collins sent out the word: The models were showing that a bomb was about to explode off California in about a week.

This bomb would come to be known as the January Fourth Storm, and it developed into a once-in-a-century weather event. To this day, no one has completely hindcasted the storm to pinpoint how it originated. The variables that spawn legitimately extreme weather events are staggeringly complex and are thus almost impossible to predict more than a week out. That's why eight or ten days before a major weather event, supercomputers owned by the European Centre for Medium Range Weather Forecasting, the U.S. Navy, and NOAA often spawn wildly divergent models. Yet from the moment that the first in a line of powerful disturbances began dumping copious snow on Mongolia, all three major models came into unexpected agreement: something resembling an ocean-size hurricane was loading up and aiming for the entire U.S. West Coast.

While most of the western seaboard prepared to batten down the hatches, Sean Collins found this storm tricky to forecast for the handful of surfers driven wild with excitement at the possibilities it augured. If the storm played out according to models, its massive swell seemed poised to arrive at roughly the same time as the raging weather. There might be giants, but from Maverick's to Cortes to Todos, a thousand-mile-wide blast of fearsome winds would probably tear the waves to shreds. Western big wave surfers watched, waited, texted, and lied about where they might, or might not, be headed.

By New Year's Day 2008, the first of three mammoth storms had wound up into a monstrous counterclockwise gyre that swallowed half the North Pacific. But unlike the early December storm, this one didn't really become turbocharged until after it had passed east of the Hawaiian Islands.

It was the kind of storm Ross Palmer Whitemarsh or the crew of the *Andrea Gail* would have recognized immediately. At the steep wall separating high- and low-pressure gradients, the upper atmosphere became a two-thousand-mile-wide tornado that accelerated the jet stream to better than 240 miles an hour. The jet slithered like an out-of-control firehose preparing to lash the coast with rain and wet snow. Behind lay a tremendously unstable and far colder air mass whose thunderheads dotted the satellite map like city-size kernels of popcorn. Surface winds that ranged from category one to three in strength linked the first and second storms just as a third began to exit the Russian coast.

On Thursday, January 3, the first low roared ashore between Washington and Oregon. At the Walla Walla airport near the epicenter, sustained winds topped fifty-five miles per hour while gusts hit seventy-eight. The barometer made a remarkable drop of 9.5 millibars over the space of three hours before bottoming out at 28.93, an all-time record low pressure for the station. By Friday morning, the chaos had extended down to the San Francisco Bay Area: Howling winds uprooted ancient oaks and littered San Francisco with debris, and a record 2.1 million customers of Pacific Gas and Electric were cast into darkness. The first fatality occurred early that morning as well. At 6:10 A.M., Rosetta "Rosi" Costello was driving down a rural road near Gold Hill, Oregon, when winds blew the top of a massive pine tree onto her car, killing her instantly. Just over two hours later, a hard-core commercial diver named Todd Estrella led the rescue of the crew of a six-hundred-ton tug that had been blown off its anchorage in San Francisco's Richardson Bay. When Estrella's own dinghy lost its mooring, he went after it. But the tiny boat capsized in 8-foot waves. Complications from hypothermia would take Estrella's life.

All told, the storm would claim at least fourteen souls.

Early Friday, January 4, the second low was preparing its assault on the coast. Nearshore buoys showed big, steep, and stormy waves—22 feet at fourteen seconds off Eureka—but the far stronger long-period energy—a great unknown—still lurked over the horizon. Mike Parsons and Greg Long studied the charts with a mingled sense of dread and fascination. Their initial thought was to line up Twiggy and Brad Gerlach to surf Maverick's or maybe Ghost Tree, but they weren't sure. The loss of the Papa buoy's wind and swell readings meant they couldn't be certain of the size and timing on the forerunner swells till they hit the California buoy. At that point, the swell would be a mere 350 miles west of San Francisco, less than a day away. They'd need to pull the trigger on where to go before that. As they wrestled over questions of wind and wave, they realized that losing the Papa buoy had set their forecasting abilities back a couple of decades.

Twiggy was in San Francisco, trading constant phone calls with Long in San Clemente. But conditions from Seattle to the Bay Area were going straight to hell. The storms would soon barrel in from Los Angeles clear down to Ensenada and Todos—one probably indistinguishable from the next. But then an interesting feature began to show on the wind models. It was probably a digital mirage, but still. "Get on a flight down here," Long told Twiggy.

The moment Long hung up the phone, it rang again. Mike Parsons had one question: "Have you seen the winds?"

Long quickly phoned Rob Brown. "I want you to be on call for Cortes."

It was at moments like this that Rob Brown wasn't sure whether he liked owning such a bad-ass boat. It provided him with a tenuous living, sure, but dammit, he was also handcuffed to it—and to these damn big wave surfers who helped pay his bills. One of these days, they were going to kill him.

Parsons had a frantic series of conversations with Sean Collins on Friday afternoon. Normally, when winds hacked the waves along the shore to shreds, those same winds rendered Cortes Bank the wildest and most unsurfable spot of them all. But the models showed this strange fissure of calm appearing off San Clemente Island between the first and second storms. On Saturday morning, it seemed possible that the winds would rage from the south until the violent pinwheel arm of the second storm passed from west to east, pulling a northwesterly blast in its wake. Parsons needed to know: If this happened, and the windfields essentially canceled one another out, would a brief window crack open at Cortes Bank?

One part of Collins loved this, all the possibilities, nuances, and difficulties a forecast like this presented. But his dear friends were preparing an Everest summit in the midst of a hundred-year blizzard. Anyone—or everyone—aboard Rob Brown's boat could die. Easily. He didn't want to be responsible for that.

"There's no doubt the swell's big enough," Collins told Parsons. "Ray Charles could call that one. But here's the thing. You're driving out *into* the biggest storm in a decade in a thirty-foot boat trying to tow a ski. You get out there, and say the wind starts to lay down, it's still going to be lumpy and crappy. If you're really lucky, it kind of cleans up. But if the front comes through early, you're screwed. It's going to be an absolute hell ride."

Brad Gerlach didn't follow the conversation too closely. All he knew was that a big storm was coming, and if something was going to happen, Parsons would call, and the surfing would *surely* take place along the immediate coast. When his phone rang on Friday at his home in LA, he was thus caught completely off guard. "I want you to be ready for Cortes," Parsons said. "It might be the biggest ever."

Gerlach had just started dating the girl he today considers the love of his life. Aleksei Archer watched as her new beau now nervously and methodically loaded seasickness patches, jackets, leashes, and his wetsuit gear into a duffel bag and grabbed his towsurfing boards. She knew Brad surfed big waves, but she was unaware of what, exactly, that meant. As Brad methodically checked his foot straps, her eyes lit up. "Foot straps!" she said. "What a good idea. Why doesn't everybody use those?"

Parsons then called an experienced surf videographer and friend named Matt Wybenga at his home in San Clemente. "Get your stuff together, we're going to Cortes tomorrow morning."

Wybenga was a bit stunned. He had been listening to tree branches snap in his yard.

"Really?" he said. "Jesus. Well, how big is it gonna be?"

"Big," Parsons said. "Bigger than anyone has ever seen."

Greg Long prides himself on being as methodical as Mike Parsons. Still, how much could you really plan for a last-minute mission into the teeth of the worst storm, well, he'd ever witnessed. They had one boat, one driver/photographer, four surfers, and two Jet Skis, and that was it. No one knew any captain besides Rob Brown who could be roused at the last minute on some damn fool crusade to ferry a cadre of thrill-seeking lunatics out to Cortes Bank. Greg did know, however, that having Steve and Rusty on board would improve their odds dramatically. He drove to his dad's house. "I want you guys to go with us," he said.

"I'm not into it," Rusty replied.

"You know what?" said Steve. "I've spent the last thirty-five years preventing accidents and shit like this *before* it happens. It's too dangerous. You're going without any kind of a safety net. You don't have another boat or any kind of a dedicated water safety team. And you know that if it's big, and you decide to get out of the boat, it's going to be really difficult to do any rescue. I'm not going to sanction this trip."

Greg nodded silently and Steve softened a bit. "Look," he said. "You're coming from the opposite end of the spectrum from me on this. You guys are the athletes and professional big wave surfers. I know it's incredibly hard to sink a Jet Ski and that Rob's boat is incredibly seaworthy. But you really need to think about what you're doing—what you're getting yourselves into."

The rejections left Greg rattled. Two of the world's best heavy watermen, his *family*, had just turned him down flat. So what the hell was he doing? He drove to West Marine in Dana Point and updated every item in a too-meager survival kit. Flares, batteries, first-aid gear, and for the first time, he laid down the money for an EPIRB—a satellite emergency transponder beacon. He studied the instructions for this electronic measure of last resort back at his dad's house.

"I'm glad to see you at least got that," Steve said.

At Mike Parsons's tidy San Clemente home, his wife, Tara, then eight months pregnant, overheard frantic speakerphone conversations all day. Mike was flustered, and she was scared. The wind was howling, and the rain was scouring their neighborhood. Normally, you didn't see the raw, ragged source of a monster swell in these placid southern latitudes, which always tended to make things seem safer than perhaps they really were.

There was no hiding from the potential danger this time. Ten feet of snow was falling in the Sierra Nevadas, and 165 miles per hour wind gusts at Lake Tahoe were creating scouring sandstorms of snow. Mountain rivers were overrunning their banks; normally bone-dry Southern California creeks were erupting into raging torrents and drowning stranded motorists. A levy broke near Reno, Nevada, sending icy water flooding through four hundred suburban homes.

Mike didn't mean to ignore Tara, but she knew the drill. She had first seen this behavior in 2004 while they were planning their wedding. She needed help with invitations, but Mike was planning a mission to Cortes. Mentally, he was just gone. *So this is what it's going to be like,* she thought.

Tara made a conscious decision that day. This was who she was marrying, and she wouldn't try to change him. She says, "When he really knows a swell's coming, that's the only thing he focuses on. Nothing else. *Nothing.* The dog won't eat. Things around him will just be . . . chaos. But that's how he gets into his mode. He blocks out any fear. I have the opposite reaction. I fixate on something, and it gets worse and worse and worse in my head, and then I get panicky."

Weather models were changing hourly. Collins was already away from his computer, chasing waves down in Mexico and coaching Parsons on the different wind forecasts by cell phone. One model showed the calm Snips was hoping for, while another—and coincidentally the one Collins typically trusted more—did not. Parsons had everything he'd need, but ominously, he carried neither an EPIRB or even walkie-talkies. Laid out instead was his own measure of last resort—a bright orange U.S. Coast Guard drysuit. It was the same head-to-toe garment worn by Alaskan fishermen, *if* they are granted the time to climb into survival gear before their boat sinks.

At around 5 A.M. on Saturday, January 5, 2008, Parsons gave Tara a long kiss, told her he loved her, promised he'd be careful, and drove toward Dana Point Harbor.

As the team converged on the harbor, they inventoried their surf paraphernalia. With the wind still howling, Brown realized it would be *way* too rough to carry both heavy Jet Skis aboard his boat. One would have to be driven. This had been done before, on the oily smooth mission of 2003, but even in the best of conditions, piloting a personal watercraft this kind of distance was exhausting. They had no idea if this was even possible. The surfers agreed to take hour and a half turns driving the Jet Ski behind the boat while cocooned in Parsons's Coast Guard survival suit. Greg Long drew the short straw and agreed to go first.

Then, once Brown's boat was in the water, one of her two engines wouldn't start. As Rob sat there cursing, a weathered old man came forth out of the pissing rain. He jiggled some wires, literally waved his hand over the wiring harness, and then grumbled, "Now try it."

The engine fired.

Long's cell phone rang. It was Grant Washburn. He'd kept detailed swell records for all his surfing life. There had never been a bigger swell. "You're gonna have a hundred feet at the Bank," he said. "A hundred feet for sure."

Parsons told Long to prepare to suffer. It was time to leave, now and under full throttle, or "we're not gonna get there in time to surf."

The phone rang in the Parsons home. Tara listened as Sean Collins's voice came in over the answering machine. The winds were not backing down and long-period energy had finally stormed the buoys. Monterey Bay had hit 33 feet at nineteen seconds. When the swell raked the Tanner buoy, it would be 23 feet at nineteen seconds—two new records. "Mike, don't go," said Collins on the machine. "It's not looking good. It's going to be crazy. Gnarly. Madness out there."

But he was too late. Six men had set out into the teeth of a mighty gale. They were tiny specks on a vast and angry ocean.

On the Jet Ski, Long did his best to stay in the lee of the *Ocean Cat*, but there was no shelter. He navigated as if he was Travis Pastrana on a moving motocross course, linking lumps and bumps in the air and occasionally driving straight into the water, his face smashing into the handlebars when a wave loomed higher than the ski would jump.

An hour went by. Long was due to change at twenty-five miles. An Olympic-level athlete waved down the boat. "I can't take it anymore," he panted.

"We've only gone fifteen miles," said Parsons. "We were going to tell you to go faster."

Long kept going, but he was delirious by the time they reached Catalina. "We're never gonna make it," a frightened Brown told Parsons. "It ain't gonna work."

Parsons switched spots with Long and straddled the ski. "I don't wanna hear it RB. We're going."

Parsons suffered mightily for around fifteen miles, but as they neared San Clemente Island, the sea calmed a bit. A few miles farther, a peek or two of sun. Soon whitecaps slithered back into the sea, and the Jet Ski's leaps into the air became infrequent. Then as they rounded San Clemente, the groundswell became a thing of wonder. Boat and ski disappeared into vast, twenty-second-long valleys. At the tops of the swells, a strange, snowy chink appeared on the horizon perhaps thirty miles distant. The white whale was breaching.

Matt Wybenga and Gerlach had both taken Dramamine, but when the boat began to plow through the long-period waves, both started battling seasickness. It was a fight Wybenga would lose.

An hour and a half later, they approached the southern edge of the Cortes Bank and were astonished. Waves were rearing up in a hundred feet of water along the *entire Bank*—across fifteen miles of empty ocean. They had never seen anything like this. No one had. Brown gave the ancient shoreline below the surface a wide berth.

Billowing explosions filled the air to their northeast above the Bishop Rock. The clanging buoy, normally well clear of the waves, was being buried—just as it had in Flame's famous photographs of the 1990 Eddie Aikau swell. Everyone was thunderstruck. Inside the broken waves stretched a murderous caldera of suffocating foam. "As far as the eye could see, it was just a huge square of white water," says Twiggy. "If you lost your guy in there, he was just *gone*. He would have been lost in that expanse, and you'd never find him. It was just so scary."

"I couldn't believe it," says Wybenga. "I've shot waves all over the world. You could count the seconds from when you saw one throw out as it fell. One thousand one, one thousand two, one thousand three, one thousand four—till it detonated. It was so loud. Then inside, this bubbling cauldron of white water. Waves were hitting and smashing into each other. It looked like death."

"But the wind was so perfect," says Gerlach. "It was just blowing offshore into these giant mountains and creating these giant, giant tubes. You'd look at it and you're like, *maybe I could ride that*. Then it would clamshell and explode, and you're like, *oh no. Oh no*."

The water still held a mogul-field of south windswell, and some of the long-period waves literally wrapped back onto themselves like a medieval sling atop Bishop Rock. This created bizarre 10-foot-high wave trains that ran sidelong into the big sets. "We had always thought, if it's 80 feet, it's going to be perfect dome bowls out there," says Parsons. "But it looked like *The Perfect Storm*. This was the first time out there when I was like, 'Is that 70 feet? 100 feet? How are we even gonna do it?' The only time I've ever been nearly as scared to ride a wave was at Jaws. But this—the consequences were just so heavy. We took forty-five minutes just to piece together—where do we ride, and how do we ride it? It was just so ominous and overwhelming."

Of course, all four surfers had been mortally scared before—plenty of times. But this was something different. It was a naked, primal fear of obliteration and nothingness; like the first Kinkipar, Archibald MacRae, Rex Bank, Ilima Kalama, and Joe Kirkwood, they felt completely inconsequential and alone among these monstrous, shaggy wrinkles in the earth's skin. Cortes Bank was huge beyond comprehension. As if to emphasize their insignificance, the possibility of actually *disappearing*—of capsizing or wiping out and not being found—was palpable. They didn't need to be caught inside and then swept out into the empty ocean. They could be lost right here. One mistake, and you'd simply float in that choking, boiling nowhere until you drowned or froze to death. But they couldn't back out. Their whale was out there. This was the hunt they'd been waiting for all their lives.

Gerlach had come this far, and somehow Parsons had always seen him through. He thought back to how unfathomable this day would have been to an eleven-year-old latchkey kid surfing sun-dappled two-footers on a Jesus board with dolphins swimming all around. Jaws, Todos, Ghost Tree—everything he'd ever seen paled compared to this. He tried to shove the fear down deep. Digging through his gear as the boat lurched back and forth brought a wave of nausea. He'd either forgotten or lost the wetsuit booty for his left foot. His front foot would not only be numb with cold, it would be loose in the foot strap. "Two wetsuits, fins, wax, surfboards, no fucking booty!" he cursed. "The biggest day of the entire fucking century and I can't find my booty."

Half hoping this might offer a way out, he gave Parsons a hard look. "Are we really doing this?" he asked.

"Yeah."

Gerlach leapt onto the Jet Ski, still cursing. "Fuck, I can't believe I went out without my booty. Fuck. Okay, it's not my day. That's okay. It doesn't have to be my day. Every day doesn't have to be my day. That's okay. I can still be a good driver, right? I can still have fun."

"You're surfing first," Parsons said.

Parsons was grappling with his own fears; life as he had known it was changing. Wondrous days of fatherhood lay just over the horizon. For the first time since the death of Mark Foo, the possibility of his own death suddenly seemed less important than the effect it would have on someone else. "I thought about Tara and our baby," he says. "You're telling yourself, you know, you can't leave her."

In a visceral way, he felt the seriousness of his responsibilities as a husband and father tugging against his big wave addiction, and yet he was in a situation where he couldn't allow himself to think about death or caution. Doubt was as dangerous as the Bank. "The thought process, the decision to go, it's really quick," he says. "And you're not thinking about dying when you're about to ride a wave—or you'd better not be. If you're timid, it's gonna be worse because you'll make decisions on waves you wouldn't normally make. You have those thoughts and you listen to them. That's the balance. For me, I think that the thrill—and whatever it is about riding waves that big—that still somehow outweighs the thought in my mind. I still know I can drown. But I still do it."

"We were going to surf the world's heaviest big wave," says Long. "We were at a point where, no matter how big the swells were—and we're still that way—we're going to go out and try to ride it."

Parsons told a moaning Matt Wybenga. "Look, I know you're sick. I know this sucks. But just shoot everything. Because it's gonna happen right now."

The surfers huddled for a brief conference: Mike would man the ski, while Brad would ride the harpoon, then Greg and Twiggy would take up the hunt next. Without radios, no one would know if anyone—even Rob and Matt in the catamaran—were in trouble. "If you don't see each other for a while, pack it up and come looking," Mike ordered. "We're all we have."

As the surfers motored off, Rob Brown had a sickening realization. He would have to edge in considerably closer to have any hope of capturing a ride. But so much energy was wrapping onto the Bishop Rock that a great dome of water had literally been piled atop its shallow plateau—it was almost like a mesa of seawater, or the convex bulge when the ocean's surface is sucked upward in the low-pressure vortex of a hurricane. The sea level simply was *higher* up ahead, and the ocean poured off the deep sides of the Bishop Rock as a wide, roiling waterfall. It seemed a bizarre contradiction of the laws of gravity and physics; Brown and Wybenga had never seen anything even remotely like it. "I was in thirty, forty feet of water," Brown says. "And it was just flooding off the edge like Niagara Falls."

"It's so hard to explain what we were seeing," says Wybenga. "It was just baffling. It seriously did look like *The Perfect Storm*. Just all this energy from the refraction building up on the rocks. It wasn't just waves coming straight at you, but from side to side."

Brown told a frightened Wybenga to hang on before plowing straight up onto the bizarre mesa like a jetboat pilot on the Colorado River. They were close in now. Maybe too close. Concentrating on the boat, Brown could pay no attention to the sick Wybenga, who was alternating between feverishly hot and bone cold. Occasionally, during the middle of a dry heave, he would be swept by a wave like a crewman on an Alaskan crab boat. Throughout, he was to yell up to Brown, both to alert him to threatening waves and so Brown would know he hadn't yet been washed overboard. High atop the swells, perfect waves were crashing down in fifty-four feet of water atop the distant Tanner Banks.

"It was *so radical looking* when you were up there on top of a really big wave," says Brown. "I don't know if the guys surfing really see it because they're concentrating so hard. But you were looking around at the entire ocean from the top of a big cliff."

In the water, Gerlach held the tow rope, floating around nervously as Parsons idled the ski, studying the waves. Like a *Pequod* crewman, Gerlach had the curious sense of being both hunter and hunted. Parsons was trying to calculate where on the reef to put Gerlach—and when. He was like a skilled trapper in the forest—using every sense and a lifetime of physical memory to spot tiny tracks, snapped twigs, a footprint in the leaves that might indicate the movement of prey. In these waves, such swirling, subtle nuances might mean life or death.

Between the lumps and the bumps, Parsons saw a wave far outside bending onto the reef. He yanked Gerlach to his feet and tracked it down. Gerlach's front foot, his left one without the booty, was sliding around as he tried to negotiate the bumps. When he let go, Gerlach knew this was the biggest wave he had ever hunted. Long and Twiggy watched in utter amazement. "Here's one of the best pro surfers I've ever seen," says Long. "Usually it's like he's snowboarding— superlow and big carves. But he was barely hanging on."

The chop was making Brad's board chatter like the skis of a Super G racer. From seven stories up, he had an instant to weigh his options. This might be the ride of his life, but the looming route to freedom was threatening to thunder closed. Were he, say, 220 pounds, the size of Laird Hamilton, he might generate enough speed to outrun it, but Gerlach barely weighed 160. His choice was to either gun it for the bottom or give up on the wave and make a sharp turn for the top, kicking out over the lip and flying into the air before it exploded so that he landed on the wave's backside.

Gerlach says, "I was going, yes, no, yes, no, yes . . . *nooooooo*."

He had only ever kicked out over the top of a handful of towsurfing waves— this was going to be one of them. As he did, though, he hit a bump, flew ass over elbows, and was embedded in the wave's lip. He paddled and kicked like Michael Phelps in a frantic effort to escape through the wave and out the back. Parsons tore in. Gerlach gasped to the surface, badly shaken.

"We're standing on the edge of life and death out here," he told Parsons.

Gerlach's near miss made it clear. They should move off Larry's Bowl and line up farther to the northeast, farther up the point—farther than they'd ever gone. They needed to straddle Bishop Rock's mammoth head, taking a position just inside the point where the seafloor began to drop off rapidly. Triangulating based on Jim Houtz's description and Mike's reckoning of the buoy location, it seems they were surfing somewhere just above the point where the *Whitney Olsen* had first tried to position *Jalisco*. The waves were even bigger up there, surely 70 to 80 feet on the faces, but they were less choppy, since the waves breaking in front of them were knocking down the cross-swells.

Now it was Long and Twiggy's turn, and they started arguing over who would surf first. This itself wasn't unusual; they were both always so amped to surf, each wanted priority. Yet after watching Gerlach's kick out, that argument was turned on its head.

"You're surfing first," said Long.

"I went first the last time, in South Africa" said Twiggy.

"No, I did," said Long. "And I know this break."

"You've never fucking surfed when it was like *this*," said Twiggy. "You go first."

"I say it's my ski and you can get fucked," said Long. "You're going first."

Twiggy took the rope and ordered Long to drop him on the shoulder of his first wave ever at the Cortes Bank—*on the shoulder*. Baker was fighting off nightmare visions of being lost in that abyss of foam. At first, it seemed like Greg had chosen well. He hurled Twig onto an endlessly long, slopey beast. But the refractions and the strange, *Jalisco*-straddled shelf atop Bishop Rock turned what seemed like a makeable wave into its opposite. After ten seconds of bliss, Twiggy was suddenly skateboarding down a 70-foot vertical rollercoaster. Senses crackled. He felt *everything*—the minute ripples on the water, the unbelievable speed, the jet of wind in his ears, and the deafening roar at his heels. The present stretched out forever, then in an instant it was over. He kicked out after a half mile, gliding up and over the shoulder of the wave to gently come to a stop on its back. A minute or so later, he suddenly couldn't catch his breath, and he started dry heaving. *What the hell is going on?* he thought.

"It's basically an adrenaline overdose," Long explains. "Twiggy had heard it could happen, and I went back and read up on it. Heavy drug users have the same thing."

"*Every wave* I had the dry heaves," Twiggy adds. "Every fucking wave."

The water was indeed smoother at the top of the point. Parsons and Long slowly round-robined their friends into a few more, playing it as safe as they could. Gerlach tried a carve or two, but man, that was scary. Better to just point it and run like hell for the exit at fifty-five miles an hour.

"It really feels a lot like flying," says Gerlach. "But then it's like riding a motorcycle over some big fucking bumps and then down a gnarly hill and then going off a big jump—but there's a monster chasing you, too. So it's not like you can do the jump and then just pull over and pop a beer. And it burns your fucking legs man. At the end of one of those waves, your legs were just *burning*."

Twiggy compares it to snowboarding, but "snowboarding, you can stop and say, 'There's a rock there and there. I'm gonna have to right, then left, then shoot straight down.' Surfing these waves, you're making all those little decisions in a fraction of a second."

In the ensuing hour, the tide drained off a bit, the wind-chop waned, and the huge refraction waves became less pronounced. The waves also began to break in a somewhat more predictable fashion. Everyone watched awestruck as a single, butter-smooth rogue peak detonated even farther outside. It was terrible and wondrous, easily better than 100 feet high. It hurled forth an arcing, almond-shaped cerulean barrel—utterly symmetrical and with perhaps six stories of hanger space inside. As it roared down the reef like a *Saturn V* rocket, the spray was surely 150, 175 feet high. "Oh my God," recalled Parsons. "It was just the biggest, the best wave I ever saw."

The swell was still building when Gerlach dropped Parsons onto a mere 65-footer. It was Mike's first ride, and it lasted nearly a minute. His second one, too. Some of the longer ones—they traveled better than a mile from start to finish and actually seemed like they would simply unfold and roll along the entire Bank without ever offering an opportunity to escape. It was hell and heaven all rolled into one.

After another half hour of working out the jitters, the surfers began to recognize the patterns in the sets: how often they arrived, where to line up with the boils and the occasionally visible buoy. Some waves were certainly hitting *something*—on the inside. As a wave rolled in, *boom*, a massive, spitting geyser would explode straight up into the air. The fear of being shoved into that violent explosion was at least somewhat subsumed by the delicious harpooner's cocktail they were now experiencing—that longed-for rush of adrenaline, dopamine, and epinephrine. They began to surf more unconsciously, intuitively.

Still, Long says, "I was just shaking. It was so far beyond anything I've ever surfed in my entire life."

The surfers moved so far up the point and became so obscured by the spray and swells that Rob Brown occasionally wondered if they hadn't simply disappeared. When either team went for a wave, it was three or four lonely, mysterious minutes before they were seen again.

This left Brown essentially on his own in the most frightful shooting and driving conditions he had ever imagined. It was a terrific struggle—keeping one eye in the viewfinder and one eye out for rogues, while pressing the shutter, changing lenses, steering, and throttling up to keep the *Ocean Cat* from being buried by wide-breaking waves. He thought longingly of the first time he had ever photographed Parsons at Cortes; then he had only nearly died *once*.

Wybenga, meanwhile, was so deliriously seasick he could hardly hold the video camera. He had tried climbing up onto the tower with Brown, clutching the outside rail like a sailor on a mast, and managed a few one-handed shots, but that only made the seasickness worse. They were so far away from the surfers that getting a good video shot was nearly impossible. Rob would call an alert to him, he'd hit record, then the surfer would disappear behind an 80-foot wave in the foreground. Then he'd throw up and flop back down on the deck, wanly wiping vomit from his chin.

From a temporary perch atop a swell, both men watched Twiggy take an endless drop down another wave, a cascading beast that through Brown's 300mm viewfinder brought a frightfully close-up look at Twiggy's desperate charge for the exit. On any other day, it would have been the biggest thing ever ridden. Yet the ride was about to be topped in the next moment. Brown saw a tiny speck racing far outside.

A monolith was lumbering up the final stair steps and standing straight up. It eclipsed anything in Parsons's experience. His ski is capable of better than sixty miles an hour, and yet Gerlach was only able to get him in by intersecting the wave at an angle. Parsons just made it through the shower of offshore spray and then over the huge hillock of the wave's backside. He looked over his left shoulder as he dropped the rope. "I just remember concentrating so hard and thinking, 'Oh shit, this is a *bomb*. Don't fall. That's all I can remember.'"

Snips bent his knees deeply to absorb the bumps and plummeted down the wave and deep into the trough. Inside swells obscured him from everyone, including a flummoxed and cursing Rob Brown. Parsons was navigating dimensions of forward, downward, and side-to-side motion like an aviator. He was also traveling faster than he ever had on a surfboard. The wave itself was surely traveling at better than fifty-five miles an hour, which meant Parsons was going ten to fifteen miles per hour faster than that. Then, quite without warning, his acceleration suddenly ceased.

"You're going as fast as you've ever gone and it feels like someone's pulling an emergency brake," he says. "It's happened on the three biggest waves of my life, and it's a crazy feeling."

It had happened on his giant wave at Jaws in 2002. Just beneath the rear of his surfboard, a shockwave had built up as the water began to boil behind his cavitating fins. Parsons crouched deeper and pointed the board straight down, expecting the cavitation to quickly abate as he slowed—like it had at Jaws. But it didn't. The rampaging wave kept moving faster than Parsons, and it began to reel him in like a fish on a line.

"He starts going *backward* up the face of the wave," says Long. "That's the only reason we could see him. The only reason he came into the frame on Rob's camera is because he was being sucked *back up*. I'd never seen anything like it—never seen anything so big in my life."

Brown clicked the shutter on his camera as the wave roared and tumbled down the point, creating another explosion of white water 150 feet high. To the edge of this maelstrom, Parsons was being sucked farther and farther up, his angle of descent climbing through sixty, seventy degrees. By the time he was pointed nearly straight down, the wave was beginning to cascade above him. Parsons began talking to himself. "It's gonna hit you, but you gotta make this. Point it. Just stay on. You can't fall. You can't die."

Then, finally, as the world was crumbling, the water resumed its normal flow around Parsons's fins. The effect was something akin to flooring a Porsche. Parsons rocketed forward and angled off toward an exit onto the wave's shoulder, still a full half mile in the distance.

Clouds burned off, and the wind died, leaving mild, bluebird conditions. Individual moments were subsumed in a deluge of adrenaline, and the session became a supercharged blur as wave after wave was hunted and slain. Gerlach found his groove and tore across several 60-footers. Eventually, though, he caught one he didn't like. It was too bumpy, and the board didn't feel right. He kicked out early and waved for Parsons. Twiggy had slung Long onto the very next wave. It seemed Long's wave would swing wide enough that Gerlach would be able to simply float over its shoulder.

Parsons circled around for a pickup, but Gerlach was fairly mesmerized watching Long. The wave was surely 75 feet tall. It was like something from another world. Parsons screamed at Gerlach to grab the rescue sled. "I'm in the straps," Gerlach said distractedly, meaning that his feet were still firmly fastened on the board. "I'll just grab the rope."

Rather than grabbing the tow handle, Gerlach clutched the rope itself, near the ski, figuring they'd make an easy low-speed cruise away from the wave, which would sweep unbroken underneath them. This way, they'd have a ringside seat to Long's behemoth. Then Gerlach took a hard look. He had badly misjudged. Long's wave was going to crush them.

Parsons pulled Gerlach up with his 165-horsepower ski. But the foam was thick, and the ski's impellor struggled to get a grip. Gerlach steadily worked his way back along the rope while holding it in a painful death grip. He wasn't going to reach the handle. The ski dug in and they rapidly accelerated.

"I looked back," he says. "The wave just looked 1,200 feet high."

Ahead of them, the ripping current and refraction waves had morphed the inside impact zone into a class five rapid filled with motocross berms. Parsons would have to clear the berms at sixty miles an hour with Gerlach hanging on behind like a mogul skier fleeing an avalanche.

As he describes it later, Gerlach contorts his face, mimicking his concentration and the blistering pain in his palms: "Mike's just pinning it, *pinning it*. There's no, 'Mike, you gotta slow down.' There's no 'Oh, I don't know what I'm gonna do.' I was like, 'Fuck I'm gonna ace this shit.' I gotta—nail—these—jumps—right—now. Here comes the first one—knees to the chest, flying, *Yeahhhh*! Here's another one, *whooooh, uhhhh*. I almost fell."

They covered several hundred yards before the wave chased them down and devoured Gerlach. Parsons made it a little farther, but he too was soon swallowed whole, despite traveling at the ski's absolute top speed—sixty-five miles an hour.

Gerlach was hurled down deep. It was black and cold, and he was down for a long time. Despite a hefty life vest, he didn't pop up. His lungs began to burn, and then he was suddenly carried back to the surface as if rolling in a column of snow. He managed a gulp of air, which saved his life, before again disappearing.

"The explosion just shot-putted me like a torpedo," Gerlach says. "I felt like a reporter in a wind-tunnel. I'm just going so fast underwater. It was roaring down there. I came up again, but I couldn't get a breath. Then it nailed me *again*. Just hit me so damn hard."

Gerlach probably spent two of the longest minutes of his life underwater.

Parsons was farther inside when he was obliterated, which was his saving grace. His life jacket and the emergency cutoff switch attached to his wrist worked. He sputtered to the surface after a half minute of violent flogging, amazed to find the ski right next to him and simply astonished that neither it nor Gerlach's lead-weighted surfboard had bashed him in the skull or slit his jugular. Then he saw Gerlach, not fifty yards away and waving. Parsons prayed as he mounted the ski, which fired up with a heavenly roar. Knowing he had only seconds, Parsons yanked his friend out before the next wave plowed through.

The surfers all retreated to a safer area, outside the impact zone, as perhaps ten more waves in the set raged past. It sank in with all of them that if Parsons hadn't reached Gerlach when he did, and if Gerlach hadn't managed to outrun the wave for a considerable distance, Twiggy's nightmare would have come true. Gerlach would have drowned in that watery caldera and been rolled so far they'd never have found him.

Naturally, the question of quitting didn't come up. Instead, they firmed up the chain of command. Whoever was driving the Jet Ski was in charge, and whoever was on the surfboard was to follow orders without hesitation. The driver was Captain Ahab.

"No more debate," Parsons added pointedly. "I say get on the sled, you get on the frickin' sled."

The weather held. By three in the afternoon, the tide began its slow rise and the swell lurched up another notch. Greg Long wanted his own Moby Dick—a wave bigger even than Parsons's and Twiggy's. Twiggy drove far up the point, and Long harpooned one right at the apex that wasn't as big, it was bigger, and neither Rob Brown nor Matt Wybenga could see a damn thing. They didn't get a shot.

"I let go of the rope and all I could do was go straight," Long says. "But you couldn't go fast enough. I just couldn't believe it. I'm going as fast as I've ever been on a surfboard, and it still felt like I was just going *backward*."

Which, in fact, he was, just as Parsons had earlier, traveling up the wave, not down it. Parsons and Gerlach gaped as Long *completely* disappeared into the white water as the lip chandeliered above him.

I'm not going to make it, Long told himself. *I'm going to get annihilated.*

Countless tons of water blasted onto his back, and he was completely blinded for what everyone figures was better than three full seconds—the longest of his life. He crouched as low as he could. "Sometimes, if it hits you like that, it'll blast you right out into the clear if you can just keep your feet in your straps," Long says.

And that's what happened. The wave literally gave birth to Greg Long. He rocketed out of its maw at what must have been seventy miles an hour.

When he kicked out, he was vibrating on another plane of existence. He shook uncontrollably and then promptly puked.

Today, Twiggy says, "That was probably the biggest wave ever ridden."

Everyone nods.

Parsons readily concedes that Long's wave was bigger than his—perhaps 80 to 85, maybe 90 feet. Not quite 100 feet, but damn close. The trouble is, with no photo, there is no way to objectively measure it. So, at the end of the day, at that year's XXL awards, and in the history books, Parsons's wave would stand as the biggest ride—ever.

The sun dropped low and clouds began to billow in from the west. The main surge of swell abated, allowing the surfers rare occasions to lay down a carve or edge into the temple of a gargantuan, spellbinding tube. Eventually, the wind stirred until it was putting a mild chop on the water. "I'm going, 'Is the day over yet?'" says Gerlach. "Jesus. So far so good. Let's get the fuck out of here." The teams decided to return to the boat. But before they did, Long and Twiggy paused in the water, just watching and taking it all in.

"We watched one of those sets way, *way* up the reef," Long says. "Easily 80 feet, just breaking in slow motion. I vividly remember how insignificant I felt—witnessing so much of nature's energy converging on a single reef in the middle of the ocean. It was just one of the most humbling, majestic things I've ever experienced."

Their day wasn't over yet. Parsons strapped on a tiny headlamp and his orange drysuit and manned the Jet Ski for the ride home. The first winds of the next approaching storm began to rake the ocean. As Brown steered the catamaran away from Cortes Bank, with Parsons in his wake, it began pissing rain and the chop returned with a vengeance. The boys huddled in the tiny cabin to ogle photos. Matt Wybenga was completely delirious. The few rides he had filmed from his seasick perch were so woozy—seasickness-inducing themselves, really—that they were unusable for anything but home movies, and to this day, only a tiny handful of people have ever seen them. They are utterly terrifying.

Meanwhile, Parsons was enjoying a new round of torture. Now that night had fallen, Mike couldn't see the swells he was battling until he rammed them head on with concussion-inducing force. "I'm saying, fuck, fuck, fuck, you guys, don't lose me back here. And when it's dark, you're always thinking you're going to hit something. I'm like, if I hit a whale, I'm just *dead*."

Parsons waved his hand over the pathetic headlamp trying to signal the boat, but Brown was fixed on the bow, his only thought to get home, while everyone else looked over his photos and decompressed over a beer.

"It's just wham, wham, wham, back there," says Parsons. "I'm pinning it so hard, like, *fuck*, why won't they slow down? Gerr's probably telling them a bunch of shit and they're all laughing."

After twenty or so minutes, the crew finally thought to look back. Parsons was gone. They stopped and drifted around for several minutes, shining the *Ocean Cat*'s powerful spotlights into the wind and rain. Parsons finally emerged, bedraggled and pissed.

"Jesus," he said. "Slow the fuck down."

At 9 P.M., cell phone bars blinked into view off Catalina Island. Bob Parsons and Sean Collins had been phoning frantically. Collins was on the verge of calling the Coast Guard. Parsons jumped into the boat to call Tara. "We did it, baby," he said.

If Parsons thought the ride home was bad, he was astonished when the team made Dana Point Harbor and the real gale roared in with gusts up to sixty miles an hour. He shuddered to think how bad that would have been out in the open ocean.

The reaction to the surf session, and to Parson's wave, across both the surfing and nonsurfing worlds was uniform disbelief.

I was terribly fortunate to be asked to cover this story for the *New York Times,* and the next day I went to Mike and Tara Parsons's tidy home and sat utterly gobsmacked alongside the exhausted surfers as we watched Brown's photos and Wybenga's video flash across the screen of Mike's Macintosh.

The first to actually break the news publicly on January 8 was, of course, Surfline. The following day, the story I wrote, titled "Surfers Defy Giant Waves Awakened by Storm," appeared in the *Times*. In no small part because of Brown's unprecedented photo of Parsons, the piece became the most emailed sports story of the week. It also became the spark that lit the fire for this book.

News outlets across the country picked up the story over the course of the next week. All had essentially the same reaction: These guys are crazy. Brave as hell, but crazy.

New York Times editorial board member Lawrence Downes wrote an essay titled "The Next Sir Edmund Hillarys: Riders on the Storm," which raised the surfers into the lofty company of history's most famous adventurer.

Surfers reacted with their own hilarious versions of the same praise. After Surfermag.com's prolific bulletin board writer Rickoray posted an entry titled "Parsons XXXXXXXXL," dozens of posts followed:

> Spoonfish: Ghost Tree was huge but that shot is cartoon like.
>
> el_calvo: These are the kind of pictures we would draw in our notebooks in school when we were kids. Never imagined that guys would actually ride waves that big.
>
> phisher222000: 2-4' Hawaiian style. Minors.
>
> Bonzer5Fin: Surflie [sic] has video. Watched it at a big surf industry office today. All the jaded corporate guys just stood there, stunned.
>
> GetShacked: I used to work on a 35ft. steel fishing boat. And we'd get tossed around in 6' seas. Doing it in the dark with 20' seas and a waverunner in tow is just simply a bad idea bordering on stupidity or recklessness. But in their case it totally paid off. Snips balls are far larger than mine.

Then, three months later, a newly minted father, his young apprentice, and an insane boat captain took the stage in Anaheim, California, to count blessings and better than $30,000 of Billabong bounty. There was no question, really. Greg Long's 53-foot wave at Todos was the Monster Paddle Award winner.

Meanwhile, the photos Rob Brown captured of Mike Parsons represented the biggest ridden wave ever captured on film, for which Parsons won a total of $15,000 and the Billabong XXL Biggest Wave Award. In 2006, Bill Sharp altered the XXL format to represent the resurgence in paddle surfing, and he changed the XXL's main prize to a $50,000 award for "Ride of the Year"—a prize only open to motion picture entries. With no video, Parsons's ride was ineligible for the top award, which went to a tow-in Teahupoo barrel by Shane Dorian that beggared description—and earned him *Surfer*'s second ever "Oh My God" cover.

In the end, Flippy Hoffman, Sean Collins, Bill Sharp and the other XXL judges were not even sure how big Parsons's wave really was. Normally you'd measure the height of the crouched surfer in his position on a wave and simply take calipers to determine the wave's height from the trough to the crest by a multiple of the surfer's height. But in the photos, there's a wave in front of Parsons that makes it impossible to see the trough of his wave. Eventually, the XXL judges determined Parsons's wave to be "70 feet plus," and Sharp sent it to Guinness to have it declared a new world record. And yet, the wave is "plus" by quite a few more feet. In my humble opinion, hidden behind the wave in the foreground, there must be at least another ten feet of slope height to the trough of Parsons's wave. This would make it roughly 80 to 85 feet high.

In the fall of 2010, I managed to corral Rob Brown and all the surfers from the January 5, 2008, mission at a breakfast joint in San Clemente called Antoine's Café. We dissected a session that each surfer still considers the heaviest surf experience of his life. Scrolling through Rob Brown's staggering photos, and Wybenga's drunken video, the air filled with ooohh's, ahh's, laughter, and profanity.

"I'd forgotten it was *this* wild out there," said Snips.

As Mike, Brad, Greg, Twiggy, and Rob reminisced, the emotions and excitement of the day returned. You could see the jones in their eyes, the longing for the hunt. "You know," Long said. "I hate to flip-flop on what I said about big wave surfing not being about the adrenaline. But we do put ourselves into a circumstance that you have just such a heightened state of awareness. You just have so much energy pouring through you. It's like no other experience. I don't want to admit I'm an adrenaline junkie. But maybe I am."

They had no doubt that they had witnessed 100-foot waves that day. The obvious question was, when would one finally be ridden?

"Everybody always asks, 'Can you ride a 100-foot wave?'" said Long. "Well, if you found one that had a user-friendly slope, you could ride one 200 feet. I still think we're so far from realizing what we're capable of when it comes to riding big waves."

"We'll need heavier boards," said Gerlach.

"The equipment has to evolve more," said Parsons, referring to his terrifying cavitation episodes. "We're bumping against the ceiling."

Everyone said the reality and, frankly, the stupidity of the mission had sunk in. Gerlach feels this is particularly true when you consider the fact that Brett Lickle nearly bled to death a month before their Cortes mission when he was slashed by an aluminum fin only a few miles off the Maui shoreline. The event left Lickle physically and emotionally scarred for life. "Those fins, they're just knives," Gerlach said.

Reminded that he had no walkie-talkie or EPIRB, Parsons said simply, "We were just idiots."

"If you talk to any accomplished captain of any vessel that's ever been out to Cortes," explained Long, "and you tell them you wanna go out there into the middle of the worst storm ever when the seas are in excess of 20 feet? They'll tell you, 'You guys are out of your fucking minds.' Which, essentially, we were. That trip could have gone either way real quick in a matter of seconds. We made headlines because we made it happen, but we could have just as easily been on every single news channel, and everybody across the world would have been saying, 'Look at these fuckin' idiots.' I still think about that regularly.

"But I mean, with everything combined that goes into surfing out there—the wave, the location, the accessibility—it really is the Everest of big wave surfing. That year, 2008, was the opportunity to reach the summit."

The one regret I heard came from Gerlach, who is still gnawed by the memory of how he backed off the first wave of the day, or what would have been the biggest wave of his life. "I totally regret not going on the wave," he said. "Other guys more fearless than me would've just gone."

Greg Long interjected: "I watched the whole thing like it was slow motion. If he'd gone down, he would have had his ass handed to him. If he didn't die, it would have set a different tone for the entire session. That's the amazing thing about surfing. It's so instinctual and in the moment. You can always sit there and ask, 'What if? Should I have?' But if you're not feeling it, if that's your natural instinct and reaction, there's a reason for it."

Gerlach considered this and nodded. "What's hard about not going is that you regret it. But I'm sitting here today. That's good, too."

Parsons reiterated this. Being a dad has definitely changed his perspective on surfing big waves. He wants to be there to watch his son, Grant, ride his own waves, and his budding little daredevil is already giving him a newfound appreciation of how Parsons's own mom and dad have felt for the last generation of their lives—watching their son lay his life on the line. And yet, while justifying the risk of Cortes Bank is more difficult, it's still not impossible. "I just can't imagine not doing it," Parsons said. "That terrifies me even more than doing it. If I missed a huge, glassy, perfect day at Cortes now, that would eat me up worse than ever."

Greg Long echoed the sentiment. "I'm fiending," he said. "Fiending to go out there. Ask anyone who's been out there and really experienced it: It's one of the most incredible places in the world, if not the most incredible. Of all the waves and all the places I've surfed in my life, without a doubt it's the most extreme. *Without a doubt.* I think of all the times I've looked around the world for waves, and all it takes is a single rock outcropping in the wrong place and, sorry, you can't surf it. That Cortes is just another one of the Channel Islands that's just not quite breaking the surface of the ocean. That it's so surfable on the right day. Everything about the place—that it came to be at all—is literally a miracle."

AFTERWORD

"The more I dive into this matter of whaling, and push
my researches up to the very spring-head of it so much
the more am I impressed with its great honourableness
and antiquity; and especially when I find so many great
demigods and heroes, prophets of all sorts, who one way
or other have shed distinction upon it, I am transported
with the reflection that I myself belong, though but
subordinately, to so emblazoned a fraternity."

—Ishmael, from Herman Melville's *Moby-Dick*, 1851

I have to hand it to Sam George. Despite working together for several years at
Surfer during the late 1990s, and despite possessing what he himself describes
as one "of the loudest voices in all of surfing," he never—as far as I know—
betrayed Flame in revealing the Cortes Bank. I was, thus, as shocked as anyone
when Swell/Surfline.com revealed the Bank to the world in January of 2001.

My obsession with the Bank took root slowly, taking a backseat to my own
addictions to storytelling and the simple need to make a living. In the years
after 2001, I continued work as an environmental editor for *Surfer* and wrote
or contributed to around two hundred stories for the *New York Times*—work
with a team of remarkable people for which I am profoundly grateful. I covered
murders, wildfires, mudslides, the X Games, Arnold Schwarzenegger's ascen-
sion to the California governorship, and Michael Jackson's freak show of a trial.
I spent humbling time with wounded Iraq veterans and people who had lost
loved ones to terrorist bombers and tsunamis. On one ill-advised assignment, I
flew with a pair of kids who repaired a battered old drug-running airplane and
became the youngest pilots to circle the globe. On another, I careened over a
thousand-foot-tall sand dune with a drunken sand rail jockey at 110 miles an
hour. On another, I sat alongside a maniacal Swedish drift-racing champion in
a 500-horsepower Dodge Viper as he slid sideways around Monterey's tortuous
Laguna Seca race course at 120 miles per hour.

I suppose that in some ways, this desire to sit alongside people who have
experienced the remarkable is its own form of high sensation-seeking. I know
I've taken some risks, but in my mind, there's a big difference between writing
about a risky behavior and actually *doing* it. Still, one result of this work was

a simple fascination with thrill seekers. Riding out to witness huge waves at Maverick's on the back of a Jet Ski with Grant Washburn is the cheater's way to experience the Almighty, and to me, it makes big wave surfing seem an even more fascinating and lunatic pursuit than it appears at a safe distance on the TV screen or from the sand.

But despite how many people still see them, most big wave surfers are not crazy. They're a small, argumentative brotherhood—complex, cerebral, raw, and damaged—who take their seemingly insane quest no less seriously than Jacques Cousteau or Sir Edmund Hillary. It's religion, passion, and science all rolled into one, and they pursue it with discipline and an unrelenting fervor.

The hook for this book was set with the barrage of deadly swells in late 2007 and early 2008. At the end of 2007 I wrote a feature for *Men's Journal* that began with the near-death experiences of Brett Lickle and Laird Hamilton on Maui and ended with a heart-wrenching day spent interviewing Peter Davi's eighteen-year-old son Jake two days before Christmas. Then, when it seemed things simply couldn't get any deadlier, the Cortes Bank left the surfing world stupefied with pictures of waves better than 70 feet high.

I began diving into what tidbits I could find on the Bank's history, hoping I might pitch a feature to *Men's Journal* or *Outside*. But the more I actually looked, the less anyone really seemed to know. Every story, even those from the most in-the-know surfers and divers, was half-filled with rumor and conjecture. The facts might easily have turned out less interesting than the rumors, but in almost every case, the actual history of the Bank was far more compelling than the whispers. If Mike Parsons and Greg Long were obsessed with riding the Bank, I became no less obsessed with telling its story.

I would spend at least a thousand man-hours in dusty libraries, on the Web, and interviewing people in person and over the phone before I finally had my first chance to see this great and wondrous ghost wave for myself. The first inkling came with a warning from Jason Murray and Greg Long, a few days before Christmas 2009.

The freshly minted winner of the Quiksilver in Memory of Eddie Aikau, Greg was first bound for Jaws, while Jason was feverishly tracking wind models and working to secure a Cortes ferry. If the winds cooperated, this Christmas Day Jaws swell would breach cleanly atop the Cortes Bank on December 27. Yet by Christmas Eve, the winds off Kinkipar seemed poised to rip the swells to shreds and the stand-down order was given.

I'd like to say I was disheartened, but in a palpable way, I was relieved. The stories of George Beronius, James Houtz, and Ilima Kalama, recently told, were resonating heavily, and actually visiting and maybe freediving the Bank, frankly, scared the hell out of me. But when the call came on the day after Christmas, there was no time to reflect. I was in Atlanta of all places, visiting family and bombing down a hill on a skateboard with my four-year-old daughter when my cell phone vibrated. "Dixie, where the hell have you been?" said Jason Murray. "It's on. We're leaving Newport Harbor at midnight."

Seven hours later, I'm at LAX, shaking hands with Greg Long, Twiggy, Mark Healey, and Flippy Hoffman's nephew Nathan Fletcher. Twiggy and Greg have thousand-yard stares. Jaws's tow-in Christmas gift was a near-death experience for both. Twiggy's hands and feet are particularly shredded, from a trip to the seafloor. "And I did a full backwards upside-down suplex in the barrel," says Greg, who only two weeks earlier won the Eddie Aikau at Waimea Bay. "Nobody's doing those anymore."

Tomorrow, though, Greg doesn't plan to tow. It's something he had discussed with me on the down-low a couple of months before. Ever since Evan Slater and John Walla stared down the gauntlet in 2001, Greg's wanted to see if it's possible to paddle into big waves on the Bank. In October 2009, he put the top secret invite out to a crew of his closest paddle surfing friends. On hearing this, I'd brought up the fact that Evan Slater faced the most frightful moment of his life paddling the Bank, and that essentially, everyone who's died surfing, has died paddling. Yet Greg, of course, had been undaunted. The obvious question then was, okay, at what point will you decide it's just too big to paddle? He'd raised an eyebrow and said, "Well, I guess that's what I'd like to find out."

Just before midnight on December 27, we meet Rusty Long, Kelly Slater, Peter Mel, and the rest of a crew of the best big wave surfers on Earth aboard a 105-foot Westport yacht, a fantastic luxury vessel somehow rustled up by Murray. The only two missing and completely incommunicado surfers are Mike Parsons and Brad Gerlach. Snips has disappeared to Baja to chase a swell. Brad and his girlfriend Alexsei have been lost in Indonesia for months. Greg settles down to a plate of steaming tamales. "I almost drowned yesterday," he says. "Now I'm sitting on a yacht eating fresh tamales and heading to Cortes. It's surreal."

A Chilean charger named Ramon Navarro has caught a cab all the way up from San Diego. His driver was an elderly guy with white hair and a beard like Gandalf who lit up when Ramon mentioned the Bank, claiming to be a fisherman who used to surf out there. "He knew all about the place," says Ramon.

Subsequent phone calls to the cab company only reveal the guy's name: Ziggy. I never hear from him.

Before we turn in, Nathan Fletcher, a quiet and intense guy I've never met before but come to like a lot, briefly recounts a tale related by his uncle Flippy of an experience just off the southern edge of San Clemente Island—a spot we're due to pass in a few hours. "You don't know what to expect out here, really," he says. "It's at the edge of the continental shelf. Anything can happen.

My uncle was out on a day and it was, like, 15 to 18 feet. All of a sudden a 100-foot wave—a rogue wave—came and they were motoring up it, and the boat went backwards over the falls. He had to jump off and swim to San Clemente Island. He said it's still the biggest wave he's ever seen."

Flippy Hoffman had not mentioned this story to me. I tell myself that we're on a big, safe, modern yacht. What could happen? At dawn, a rogue wave seems unlikely but not out of the question. Strong high pressure leaves a gauzy haze on the horizon, making it impossible to tell where sky ends and sea begins. The mirror-smooth water teems with tiny, hookfinned dolphins. We're still several miles off Bishop Rock, making a southerly approach and passing right along the top of the Cortes Bank with nary a wave in sight. Still, the ocean is oddly woozy, just as Sharp had described on his first mission out here twenty years ago. In fact, he says, the morning's almost a carbon copy.

Half an hour later, my heart leaps at a strange apparition. I lift my binoculars to see the solid lines of the new long-period swell. The first wave rises majestically beneath diffuse morning sunlight, a perfect A-frame peak. Straight out is the CB-1 buoy, waving back and forth. A sea lion rockets out of the water and climbs aboard. His cousins bark angrily for a moment before returning to their languid naps. Ten minutes later, we're close enough to hear a wave shatter the morning quiet. I feel the strangest déjà vu I've ever experienced—a homecoming to a nowhere I've never been. On seeing the waves and actually being out here, the fear and trepidation of the previous few days is just *instantly* subsumed. All I want is to get out there and see it up close.

A hardy young fisherman named Nate Perez has ferried the surfers out to the middle peak, clad in nothing but a T-shirt and a pair of shorts. I soon jump onto his ski, heart in my throat, wearing jeans and a windbreaker and clutching my unprotected Nikon. A hundred yards out, I think of Joe Kirkwood and his fur boots, and sense this is a mistake. I know it's dangerous, and really, I'm scared half to death. But for some reason, I don't really care and I don't have complete control over my decision to jump on the ski. It's a strange, almost out-of-body-sensation. I've never been hypnotized before—by humans or the ocean—but I think that's what's going on.

The waves aren't giants by Cortes standards, which actually makes them perfect for an inaugural paddle surfing mission. The very biggest are perhaps twenty-five feet from top to bottom. Still, they're just incredibly powerful— shifty humpbacked bulges of frictionless energy that glide through the ocean with preternatural speed. Some hit far up on the north peak, capping over and rolling down the line like mutant runners at Trestles. Others rise into explosive, cone-shaped wedges that Kelly Slater says remind him of Sebastian Inlet, Florida, on steroids. These cones make it obvious why Captain T. P. Cropper once thought he was above a volcano. Other waves shift a little farther toward us and jack onto Larry's Bowl in deep, barreling slabs. Out off the middle peak, the surfers hoot. Greg knifes cleanly into a butter-smooth twenty footer. He draws a beautiful turn and angles for the exit—perhaps the first wave paddled into out here since Bill Sharp's last wave back in 1990.

A seemingly makeable wave then sweeps up Ramon Navarro, but it scoops him up far faster than he can paddle. He falls into the air and slides down the face on his back like he's at a waterslide and is absolutely murdered. Again, I feel this strangely clearheaded and fearless fascination. I mutter to myself, "Wow, what must *that* have been like? I wonder if he's gonna go over the ship. Jesus, how hard would I be panicking right now?" Every so often a sea lion surfaces and slings a yellowtail into the air among the surfers. Mark Healey, a man just off a trip to Guadalupe Island where he actually *rode* on the back of a great white, scares the shit out of everyone with jokes about the sea lion carcass floating out at the edge of the lineup. I had hoped to paddle out on a spare board and snorkel a bit in the zone well inside the broken waves. The poor sea lion has convinced me that's not going to happen. I'm feeling foolishly brave, but not bravely foolish.

Perez prowls outside beyond the breakers. At every other big wave spot I've ever seen, there's this constant background roar from breaking waves hitting the shore. Not here. The tiniest whisper of south wind is enough to blow a huge cascade of spray off the hulking backs of the swells. This rainstorm is the only thing you hear until the tremor when the wave folds over. It's a profound silence punctuated by rain showers, hoots, and explosions. Then when the explosions have abated, and the set has passed, it's again still and quiet as a tomb. Utterly surreal.

When Perez idles back over inside the lineup, and kills the engine, fear is mingled with sheer wonder. We can just see straight down through air-clear water into kelp that waves to and fro like a mermaid's hair. Perez warns me to keep an eye out for strands that might choke the impellor. Just to our left, the water churns and swirls—thousands, maybe millions of shimmering menhaden swim in a tight tornadic vortex. A sea lion pops up with a loud snort, no doubt snacking on the fish, and startles the crap out of us. We drift in the current above the forest, and eventually catch a glimpse of white, almost like a dusting of snow. With shock, I realize we're actually staring at the sandy bottom, or really, the top, maybe no more than twenty feet down. I can see Archibald MacRae's rock.

The sand is intermingled with black stone and forearm-thick trunks of of kelp. Golden garibaldi the size of dictionaries weave all around. A big bat ray soars just off the stern like a spotted pterodactyl. The sunlight throws out the brilliant rays of a divine disco ball.

So complete is our distraction that we fail to notice when the water is drawn down off the reef. When we finally look up from the pit of Larry's Bowl, the view is equal parts dream and nightmare. The wave is only two, maybe two and a half stories tall, but I've never stared down anything remotely this big from the actual firing line. It's beautiful and expansive and I sort of feel like I'm standing before the kelp forest tank at the Monterey Bay aquarium, but there's no glass. Like Kirkwood, I see fish, yellowtail and striped mackerel plainly visible. The world seems to move in slow motion—like it's being shot at 120 frames per second. It would be easy to just stare straight at this seemingly impossible thing until it kills you. Indeed, when Nate first keys the starter, nothing happens. He hits it again and the engine sputters to life. "I hope we didn't suck in any kelp," he says as we're drawn onto the halfpipe base of the wave's face. We roar out and I look over my shoulder as the wave gnashes its teeth, cursing at the one that got away. I later wonder if this odd, slowed-down mingling of naked fear and slackjawed astonishment is what it's like to stand before an onrushing tsunami, or maybe, on the deck of a sinking *Jalisco*.

We pull up to the yacht and time speeds up. The only time I've ever felt as high was eight months earlier as I waited an interminable few seconds to hear my newborn son take his first coughing breaths after a dangerous delivery. The water, everything around us, seems to sparkle with a supernatural light. I'm delirious. *Spun out.* "How was it?" asks one of the crew.

"I'm just celebrating my survival," says Nate.

At nightfall, the yacht veers off towing five skis, the captain taking a similar route over the Bank that brought us here, through waters between 70 and 150 feet deep. Pete Mel has stroked into several bombs and is the only one who was out on the 2001 expedition. I ask him about the difference. "Towsurfing, you're just waiting for a wave to come and positioning yourself wherever you want," he says. "You're not sitting out there trying to read the lineup and the boils. Today you had a completely different mindset. It was actually way more intense."

On the dark stern deck, Greg and Rusty high-five and relax over a couple of beers. Off the port quarter, Rusty sees a dark shadow. "What the hell is that?" he asks Greg.

A black hillock falls forward and rampages toward the boat. Long yells up to the bridge. But the captain has already seen the rogue. Everyone feels the yacht accelerate wildly. It makes a perilous downhill plunge and then seesaws back up as the wave passes beneath us.

Three of the skis have snapped off the towline. The spooked captain pulls out off the edge of the mesa while Long and Ramon Navarro grab tiny flash-lights and walkie-talkies, wetsuit up, and head out. They zip back and forth between the blinking buoy and the lit-up yacht for half an hour, their flashlights twinkling, as they dodge lobster trap marker buoys. Nothing. Finally, a wave rears up and Long is carried high up a twenty-five-foot face in pitch darkness. He catches a glint off in the distance and roars off to reel in the skis. Back aboard the boat, he considers what he's seen. "If we'd been running fifty yards farther over," he says, "we would have yardsaled a super yacht."

The beer flows and the conversation soon nervously, and then hilariously, resumes.

November 1, 2010.

It would be not quite a year before I'd again see waves above the Bank. At first, the forecast looks like a repeat of January 2008, but the storm takes a more northerly track and the swell loses some of its punch between Hawaii and the mainland. Still, it's going to be way, way bigger than it was eleven months earlier. I call Jim Houtz to ask if he'd maybe like to climb aboard the boat with a group of big wave surfers whose obsession with the Bank at least matches his own. "Are you kidding?" he says. "Sign me up."

Bill Sharp and Jim Houtz have towed a trio of skis down to San Diego and we meet Captain Scott Meisel and the hilarious crew of a ninety-foot sportfisher called *Condor*. The engine throbs and Bill Sharp takes a seat alongside Houtz. "Now tell me this story," he says, "I wanna hear about the fiasco on the *Jalisco*."

Houtz thumbs open a scrapbook and is soon encircled by a small troupe of wonderstruck watermen.

A little after midnight, I retire to a tiny bunk. The swells soon reach a near carbon copy of January 2001's 15 feet at twenty seconds. I imagine that's why I have terrible, awful dreams of a ceaseless, rolling earthquake. When I'm finally overtaken by a world-ending tsunami, I wake with a jolt, bang my head on the bunk, and stumble out onto the dark bow for some air. On the bigger swells, *Condor* lunges skyward and reverberates with a giant shudder as she drops into a black hole. The hour would put us a short distance off San Clemente Island, and of course, I can't help but think of Nathan Fletcher's story of his uncle Flippy. But again, the sense of fear is again replaced by wonder. The ocean is a living thing and we rise and fall on her deep breaths.

There's little comfort off Bishop Rock five hours later. Captain Scott assures us we're in 180 feet of water, but a packing foam plant might as well have exploded out here for all the froth. Rob Brown approaches at dawn, ferrying Jason Murray, Greg Long, Mike Parsons, Mark Healey, Shane Dorian, Ian Walsh, and Jeff Clark. All are bleary eyed from the previous day at Maverick's where they would learn of the death of their friend Andy Irons. Out on the eastern horizon, daylight reveals the summit of Kinkipar. It's the first time even

Snips has ever seen the island from Bishop Rock. If you were standing on the summit of Mount Thirst right now, you'd be granted a dazzling view from Catalina to Santa Barbara to San Diego clear out to this bizarre patch of open-ocean white water.

The waves are just fricking *massive*. Bigger by an order of magnitude than anything I've ever seen—ever imagined. And yet they're only 50, maybe rarely 60 feet high. Way, way off in the distance a monstrous and completely unexplored lefthander tears along Bishop Rock's eastern edge. Spinning like a fifty-foot-tall Tasmanian devil, we reckon it must be a mile long. The surfers stuff flotation pads beneath their wet suits that make them look like bulging action figures (which, of course, they are), and four skis peel away looking for all the world like X-wing fighters attacking the Death Star. Sharp and I jump onto Rob Brown's boat. He brings us in close to Larry's Bowl and orders us to keep an eye on our port flank for rogues. The waves are much steeper here than out by the *Condor*. They loom far above us, threatening to break. Rising to the top of a swell, I recall Matt Wybenga's description of the boiling zone of damnation inside the broken waves. It looks like death. I'm sickened, but there's a part of me that really, *really* wants Rob to pull in even closer.

From a position that Brown calls "relatively safe," but feels anything but, I make a few new observations. One: When it's massive out here, the cacophony produced by a steady progression of 40- to 60- foot waves is not only that of the thunder, but of the gales they create as the air struggles to get out of their way. The waves don't just roar, they *howl*. Two: When Greg Long yanks Snips onto a giant, he barrels down the line making these *beautiful* turns. Suddenly, though, the water explodes 150 feet into the air behind him like a depth charge—a roiling cumulonimbus cloud sprinkled with rainbow dust that Parsons's completely unaware of. A wonderstruck Jim Houtz reckons maybe the wave bashed some part of his ship. Three: When I later replay the shaky footage from my video recorder, a quick, blurry pan reveals something black and immovable way out there in the white water—right where MacRae's pinnacle should be—right where the *Jalisco* met her doom. I show it to Sean Collins and Bill Sharp. Both think it's just the face of a wave. Yet Rob Brown disagrees: "I see a big rock," he says.

A line of fog looms on the northern horizon. Minutes later, it buries us like a sandstorm. Motoring back to the mother ship, Sharp muses, "How would you like to be cruising along out here in your clipper ship in the eighteen hundreds, and you're in a fog bank like this. Suddenly you come upon this place and you say, 'Hmmm, what's that? Sounds like cannon fire.' Oops."

A couple of hours of laughter and incredible fishing pass back aboard the *Condor*. Then as suddenly as the fog appears, it's gone. "Towsurfing's too easy," says Healey, after reeling in his second yellowtail. "Let's paddle."

I didn't expect they'd actually paddle today. I mean, the swell has dropped a bit, but the breeze has risen and the waves are still damn big. Long, Healey, Ian Walsh, and Shane Dorian are not even sure if it's possible to paddle in out here at this size, but they reckon that the westerly wind at their backs and the resulting chop might allow them to sort of "chipshot" into a few waves. Still, the playing field is so vast—bigger than any other big wave spot. There's a sort of resigned understanding that they're going to be absolutely annihilated, but they still man their harpoons.

The paddlers will soon agree with our observation from the boat. The waves look way bigger with a person set against them. "It's so eerie out there," says Long. "It's impossible to figure out. You just have to pick a spot and hope it will come to you and then hope you don't have a 60-foot fucking white water come and take you out. It happened to Healey and me. You paddle over the top of one, and there's another one, just breaking fifty, a hundred yards outside of you. It's like, 'What do you do?' Relax, take a couple of deep breaths. Then when it hits you, it hits you hard."

Healey is more succinct. "I was just thinking to myself, so this is what it feels like when you don't stand a chance."

I can't help but recall Healey's words four months later, when a paddle session in waves about this size at Maverick's will widow the wife of Sion Milosky, and leave his two beautiful young daughters without a dad. Nathan Fletcher will discover his best friend's body. A similar brush with mortality nearly widowed Shane Dorian's wife a year earlier at Maverick's. While trapped on the bottom, Dorian decided something had to change.

Today, when Dorian is pushed down very, very deep onto the Bishop Rock and is quite certain he doesn't want to be down any longer, he pulls a little ripcord on his shoulder. This pierces the casing of a metal CO_2 cartridge, instantly inflating an air bladder woven into his wet suit—the same principle behind an airline life vest. He rockets up through perhaps forty feet of foam and returns to the boat sporting a huge hunchback but quite alive. This simple invention might have just saved his life. As far as Mike Parsons is concerned, if he and his buddies are going to keep risking their necks out here, Shane Dorian has just defined the future.

The next morning, Long, Parsons, and I ride back up to San Clemente with Jim Houtz. Long is still decompressing from the paddle session and the throttling he and his friends have survived. When he considers all the stories—Jim Houtz and all the other people the Bank has nearly, but not quite killed—the fact that he's been surfing over a shipwreck site—he senses a touch of the divine at work. Someone, or something, has been looking out for scores of people through the years on Bishop Rock. "When you're paddling all alone out there, when you *really* look at the place and feel its immensity," he says, "you can't just help but feel that there's something so much greater—so much more significant at work than you."

Houtz agrees. "What happened to us that day in 1966—the way those waves just came out of nowhere, the fact that everyone survived—it really was hand of God stuff."

To James Houtz and Joe Kirkwood, the Bank held the promise of a resurrected Atlantis, a Nietzschian citadel. Mel Fisher and Ilima Kalama sought the bounty of a sunken Treasure Island. For Greg Long and Mike Parsons, the Cortes Bank is a sort of supernatural Everest—an ethereal, momentary mountain that holds the promise of growing higher with each expedition. For all of us, this ghost wave pulls like Melville's whale, an enigmatic monster that lures you out for the hunt—and nearly kills you every time.

I've been profoundly humbled by the Bank and deeply honored to tell its stories.

ENDNOTES

I must note a great deal of editorial assistance from two people. The first is my mother, a supremely talented editor and researcher named Gloria Ricks Taylor. The second is talented and hopeful surf journalist and U.S. Air Force pilot, Captain Steve Stampley. Both went so far above and beyond in helping me to chase down facts, copy edit, and fit this book together that it's almost inconceivable.

CHAPTER 1

1. Recollections came through a few years of in-person conversations with Bill Sharp and Sam George, and telephone interviews with Bill, Sam, and George Hulse in 2010.

2. Background on Larry "Flame" Moore came from Sharp, Sam George, Mike Parsons, and Sean Collins; from interviews with his wife Candace during October 2009; and a lengthy but unpublished interview Sharp conducted with Flame shortly after the 2001 mission to Cortes Bank. I had only one occasion to meet Flame, when I went to the *Surfing* offices to pick up a photograph for use on Surfermag.com in 1998. He was friendly, energetic, and *very* explicit over the photo's care and feeding.

3. I interviewed Philip "Flippy" Hoffman by telephone in August 2009. I had hoped to talk in person about the rogue wave described by Flippy's great-nephew Nathan Fletcher, but unfortunately, Flippy passed away at the age of eighty, on November 15, 2010. I have been unable to confirm his recollection that the navy once blew the top off Bishop Rock.

4. The collision of the USS *Enterprise* was briefly described in "Nuclear Carrier Enterprise Hits Reef Off San Diego," *Los Angeles Times*, November 4, 1985.

5. Nick Carroll, ed., *30 Years of Flame—California's Legendary Surf Photographer* (San Clemente, California: *Surfing* Magazine, 2005). Carroll's excellent book was immeasurably helpful in painting a picture of Flame's life.

CHAPTER 2

1. The spark for this chapter was lit by intriguing interviews with Dr. Rikk Kvitek, director of the Seafloor Mapping Division at California State University, Monterey Bay, and his colleague Dr. Gary Greene. I should also note the assistance of NOAA fisheries biologist John Butler in relaying the gloomy state of the Bank's abalone fishery.

Kvitek's mapping mission to the Bank was a typically wild ride. First, his remote-operated camera was ensnared in the propeller of their research vessel. "Then at 2 a.m., all the data stopped coming onto our screen," he said. "And sparks and flames started shooting out of the smokestack—we had a flue fire. Then I looked off the side and saw the thickest school of mackerel I've ever seen in my life—millions and millions thrashing everywhere. I went to pull up the sonar but it had been snapped off. That's why we don't have the best data set. We ran into this feeding frenzy and the best indication is that the sonar clipped a feeding whale."

Kvitek's hypersensitive sonar thus fired blanks along portions of Cortes's shoalest waters, but he was still troubled and amazed at his scans. First, largely illegal commercial overfishing has nearly eradicated the Bank's abalone. Then there was the strange seafloor. At Tanner Bank, Kvitek imagined a lagoon, 250 or so feet deep, ringed with low rock hills, where canoe-bound Native Americans once hurled spears and cast nets. "Then what you see at Cortes are old, historic shorelines," he said.

Dr. Greene said that Cortes and Tanner were heaved to the surface by tectonically tortured basalts in a process that may be ongoing. At Bishop Rock and Cortes's nine-fathom shoal, softer sandstone has been scoured away to reveal a hard, black basaltic heart—a formation called a "buried hill." "These are volcanic rocks, twenty million years old," said Greene. "They're relatively young in geologic time, and they're exotic in that we're not sure exactly where they came from."

2. The following sources were also invaluable:

A. Chuck Graham, "Chumash Tumol Makes 22-Mile Crossing," *Canoe and Kayak Magazine* Web exclusive (http://www.canoekayak.com/canoe/tomolcrossingchanelislands).

B. Roberta Reyes Cordero, "Our Ancestors' Gift Across Time: A Story of Indigenous Maritime Culture Resurgence," *News from Native California* 11, no. 3 (Spring 1998): (http://channelislands.noaa.gov/drop_down/chumash.html).

C. Charles Frederick Holder and Juan Rodríguez Cabrillo, *The Channel Islands of California: A Book for the Angler, Sportsman, and Tourist* (Chicago: A.C. McClurg and Company, 1910). The above book also refers to Fray Geronimo de Zarate Salmeron, *Relaciones: An Account of Things Seen and Learned by Father Jeronimo de Zarate Salmeron from the Year 1538 to Year 1626*, trans. by Alicia Ronstadt Milich (Albuquerque: Horn & Wallace Publishers, reprinted, 1966).

D. L. Mark Raab, Jim Cassidy, and Andrew Yatsko, *California Maritime Archaeology: A San Clemente Island Perspective* (Lanham, MD: Alta Mira Press, 2009).

E. J. E. Holzman, "The Submarine Geology of Cortes and Tanner Banks," *Journal of Sedimentary Research* 22 (1952).

F. Reports published in the *Pacific Coast Archaeological Society Quarterly*. Most are available on their Web site, www.pcas.org. The following were especially fascinating:

(1) Paul Porcasi, Judith Porcasi, and Collin O'Neill, "Early Holocene Coastlines of the California Bight," vol. 35, no. 2 (Spring/Summer 1999).

(2) Andrew Yatsko, "Of Marine Terraces and Sand Dunes: The Landscape of San Clemente Island,"
vol. 36, no. 1 (Winter 2000).

(3) Ellen T. Hardy, "Religious Aspects of the Material Remains from San Clemente Island," ibid.

(4) Clement W. Meighan, "Overview of the Archaeology of San Clemente Island, California," ibid.

(5) Roy A. Salis, "The Prehistoric Fishery of San Clemente Island," vol. 36, nos. 1 and 2 (Winter and Spring 2000).

(6) Michele D. Titus and Phillip L. Walker, "Skeletal Remains from San Clemente Island," vol. 36, no. 2 (Spring 2000).

G. I also conducted telephone interviews with Paul and Judith Porcasi and Andrew Yatsko. The Porcasis alerted me to the existence of pygmy mammoths and *Chendytes lawi*, the great flightless duck. Yatsko pointed out that the Kinkipar might have gone to Cortes and Tanner Islands to hunt during strong El Niño events—whose warm water can decimate local fish and mammal populations. "They were well maritime adapted," he said. "The notion of people on those outer islands during the early Holocene period is not at all out of line." Yatsko also pointed out that "Gabriellino (natives to San Clemente and other islands) culture declined so quickly after European contact that it remains little more than a cipher to anthropologists."

H. *Island of the Blue Dolphins* by Scott O'Dell (New York, Yearling. First edition, 1961, reprinted, 1987) is a fascinating, fictionalized account of the legendary "Lost Woman" of San Nicolas Island and a must-read for anyone with an interest in the last days of California's island Indian culture.

CHAPTER 3

1. The reference to the *Santa Rosa* shipwreck is provided on the first page of the appendix of the 1981 edition of *Shipwrecks of the Pacific Coast* by James A. Gibbs (Hillsboro, Oregon: Binford and Mort, 1981). Interestingly, the first printing of this book (1957) makes no such mention. Perhaps Gibbs relied on Mel Fisher's 1957 determination of the ship's location.

2. From John Potter, *The Treasure Diver's Guide* (Hobe Sound, Florida: Florida Classics Library, 1988): "There are several published accounts that a Spanish galleon, carrying some gold, sank at the outer point of Cortez Bank in 1717. This ship was reported to have struck a 15-foot deep shoal now called Bishop's Rock . . ."

3. Logs and insight from the *Constitution* were provided by a Rebecca Parmer, an archivist with the USS Constitution Museum in Philadelphia.

4. The writings of James Alden and fellow Coast Surveyors like Archibald MacRae are published in lengthy annual books titled *Report to the Superintendent of the Coast Survey Showing the Progress of the Survey During the Year 18__*. These books were digitized by NOAA as images (http://docs.lib.noaa.gov/ rescue/cgs/data_rescue_cgs_annual_reports.html). The 1855 report contains Archibald MacRae's discovery of what would become known as Bishop Rock. The finding was reported in "Dangerous Rock on the Coast of California," *New York Times*, November 3, 1855.

5. I became aware of MacRae's death when reading the Coast Survey report of 1856, when Alden wrote of "the untimely death of that intelligent and energetic officer." NOAA historian Albert "Skip" Theberge confirmed it a suicide and suggested I read the Spring/Summer 2006 issue of *Mains'l Haul*, the journal of the Maritime Museum of San Diego. The issue is devoted to the journal of Coast Surveyor Philip C. Johnson, who served briefly under MacRae on what it seems was MacRae's second journey to Cortes Bank (I cannot definitively state that MacRae went to Cortes twice, only that reports seem to indicate so). On November 18, 1855, Johnson wrote, "Lieutenant MacRae committed suicide by blowing his brains out in the cabin of the *Ewing*. He was buried in cemetery on the 19th." Further queries revealed that MacRae was a member of Wilmington, North Carolina's most prominent family. (His great, great, great nephew Hugh MacRae Jr. is today an avid surfer). Eric Kozen, caretaker of Wilmington's Oakdale cemetery showed me his gravesite. (If MacRae's indeed buried there, he was pickled in whiskey and shipped home). Beverly Tetterton, special collections librarian at the New Hanover Public Library in Wilmington, made me aware of MacRae's letters at Duke University. At Duke, research librarian Arthur "Mitch" Fraas provided inestimable help in locating and sifting through some four thousand pages of MacRae family documents. Google revealed MacRae's lengthy account of his days in Chile in *The U.S. Naval Astronomical Expedition to the Southern Hemisphere During the Years 1849–'50–'51–'52*. MacRae's writings reveal a fiercely intelligent and witty raconteur, and are ultimately heartbreaking. They include the following:

In the Andes: *I have rarely passed so uncomfortable a night, nor one, at the same time, more impressive. My face and hands were blistered by the sun and chapped by the cold winds to such an extent as to produce fever, and I found it impossible to sleep . . . We made our fires at nightfall with mules' dung—the best fuel to be had; and as the wind was strong in squalls, our stew was pretty well seasoned with the ashes. These, however, are things to which one becomes accustomed . . .*

Letter home from Santiago: *"Sue" asked my opinion as to whether a fan she had was large enough to go to a ball with. On examining it, I discovered that it was a little broke, and pretending to think that I had broken it, I insisted on carrying it and having it repaired. To this, she consented and accordingly I carried it off. But instead of having it fixed, I bought a new one which cost me three dollars and sent it—as a reason that I could find no one to repair the other. A few nights after being there alone, Susan thanked me one night with such a sweet confession, showing at the same time a gentlemen's ring which had not even the jewelers mark off it, that I felt most infernally spooney* (author's note: enamored in a silly, or sentimental way) *but withal a little scared. I had an idea that ring was destined for me and knew that if it were once given, all would be over as far as my will was concerned . . .*

6. MacRae's delivering news of the Mexican-American War was recounted on page 287 in *The Works of Hubert Howe Bancroft: History of California, 1884-90* by Hubert Howe Bancroft (Whitefish, MT: Kessinger Publishing, 2007). Former New Jersey Governor Rodman M. Price, who served under MacRae aboard *Cyane*, also recounts MacRae's work as a spy.

7. The claiming of Bishop Rock by the United States by Commander William Adger Moffett appeared in "We Get More Territory," *The New York Times*, October 16, 1911.

8. The timeline of the SS *Bishop* and numerous articles that would question the *Bishop's* Cortes Bank collision would have been impossible to find without diver and treasure hunter Steve Lawson, whose discoveries included archival stories from the *Daily Alta*, *The New York Times*, and "About a Rock—and a Bishop," *Mains'l Haul* 5, no. 2 (1966). To learn more about Lawson, read *Lost Below: The Southwest's Most Intriguing Shipwrecks, Sunken Aircraft, Submerged Ruins and Undersea Treasures* by David Finnern (Pearl Publishing, Monterey, 2009).

9. Another source of Coast Survey lore who paints a vivid picture belowdecks is writer Kenneth Lifshitz, who has been penning a novel based on James Alden, George Davidson, and the Coast Survey he calls *Monoville*. He's looking for a publisher.

10. Early twentieth century encounters included Maude Pilkington Lukens, "Road Mapping Our Sea Coast," *Los Angeles Times*, February 22, 1925; George Wycherly Kirkman, "The Lost Islands," *Los Angeles Times*, October 24, 1926.

11. The *El Capitan* collision: "12 Fishermen Saved as Two Boats Collide," *Long Beach Press-Telegram*, November 18, 1952.

12. I interviewed Mel Fisher in around 1989 as an eager young writer for a Myrtle Beach weekly called *Hot Times*, at small museum where you could see booty from the *Atocha*. Other Fisher stories included: Lee Bastajian, "Divers Will Hunt Undersea Fortune," *Los Angeles Times*, July 29, 1956; George Beronius, "Sunken Treasure! Shout Lures 23 on Sea Search," *Los Angeles Times*, January 14, 1957; and Eugene Lyon, "Atocha Tragic Treasure Galleon of the Florida Keys," *National Geographic*, June 1976.

13. Concerning the many "explorer-divers" of the Bank: I interviewed Ilima Kalama by phone in August 2009; Harrison Ealey at his home in Oceanside, California, in October 2009; and Rex Bank in his Long Beach home in October 2010. Ilima Kalama wrote the liner notes to The Ventures' album *Surfing* in 1963 at the age of twenty.

14. Concerning the USS *Enterprise*: Some years ago, Bill Sharp located a U.S. Navy study titled "Bishop Rock Dead Ahead: The Grounding of the USS Enterprise," by Karlene H. Roberts. My descriptions were based on Roberts's study and a telephone conversation and email correspondences with Rear Admiral Robert Leuschner. I'm hugely grateful for the admiral's detailed and honest attention to my queries.

CHAPTER 4

1. When I dove into the tales of plans to colonize the Cortes Bank, dates, locations, and facts varied wildly and details of the people who planned to erect this nation on the half shell were maddeningly scant. The true nature of this epic was first revealed through articles stored on genuine microfilm, far from the all-seeing eyes of Google. The first mention appeared on page 11 of the *Pasadena Independent* on Halloween of 1966 (Hal D. Stewart, "Pair Planning Island Nation off San Diego,"). I then located a number of other stories in the Web archives of the *San Diego Union Tribune* and the *Los Angeles Times* that gave names. Yet, for some time, I was unable to locate anyone.

I learned per the *Los Angeles Times* that Abalonia partner Robert Lynch had died in 1997. Then one day I received an anonymously sent package. The folder bore a sixty-page manuscript with Joe Kirkwood's incredible account. It seems he penned the story around 1967 and then sent it to *Sports Illustrated*. But *S.I.* rejected the famed golfer's story and perhaps Kirkwood never farmed it to anyone else. I also received photos, a short film clip of *Jalisco* being towed under the Golden Gate Bridge, legal opinions, correspondence between several "Abalonians," and a study titled "A Plan for an Island State," for Bellevue, Washington–based Cortez Development Corporation. The study stated that an American engineer named Edward M. deSarra had spent more than $250,000 to plan four islands atop Cortes Bank: Taluga (2.3 acres), Aurora (26 acres), Triana (102 acres), and Bonaventura (48 acres). There would be an airport, parks, pools, putting greens, a primary school, a high school, a yacht club, and government buildings. I cannot sufficiently express my gratitude to whomever sent these documents.

Then came another break. One day in mid-October 2009, a voicemail: "Umm Mr. Dixon, this is Jim Houtz. I'm one of the guys who was aboard the *Jalisco*. Give me a call."

I managed to verify portions of both Kirkwood and Houtz's accounts through news stories. Then in March 2010, I reached *Whitney Olsen* crewman Louis Ribeiro, who said that Houtz's version of events seemed to follow his recollections. There was also record of the Coast Guard's rescue of Kirkwood and his friend Dick Hall aboard *Sallytender* in the National Archives.

2. "'Joe Palooka' wins $50 Million Purse" by Mitchell Smyth, *Toronto Star,* August 30, 1987, showed Kirkwood selling a golf course on Kauai for $50 million. "Now that he's leaving the Hawaiian hideaway," Smyth wrote, "Kirkwood is a little sad . . . 'I have a mile of beach here—it's considered the prettiest mile of beach on the Hawaiian islands.' But there'll always be his golf to keep him happy. The game is also the basis for his deep belief in fair play and consideration for others. 'It's a game where if you cheat you are only cheating yourself. What better rule to live by?'"

3. Eventually I found that Abalonian Bruce McMahan still existed. Web sites exist for several organizations, including a hedge fund, a supercar company, and a philanthropy (http://www.brucemcmahan biography.com/). Emails and phone calls to a PR firm for McMahan yielded promises that calls would be returned, but none ever were. Then I stumbled upon a lead feature (Kelly Cramer, "Daddy's Girl," *Village Voice*, September 26, 2006). Cramer reported on court records indicating that in 2005, McMahan settled a lawsuit brought by a daughter born out of wedlock. The suit had claimed psychological damages in the wake of an alleged affair and even a wedding between McMahan and his daughter. True or not, Jim Houtz put it best when he said, "That's probably why you didn't hear from him."

4. Photos of the shipwreck were taken by Daniel Bresler, who had been hired by Kirkwood and then sold them to the Associated Press. I reached his son by phone in Los Angeles who told me his father's own story—including life as a marine combat photographer.

5. Among the other sources:

A. Staff, "New Nation May Rise 120 Miles From Coast," *Los Angeles Times*, October 14, 1966.

B. Harold Keen, "Promoters of Abalone Ship Plan May Face Federal Prosecution," *Los Angeles Times*, November 17, 1966.

C. "New Island Plan Pushed by Kirkwood," *Los Angeles Times*, November 29, 1966.

D. Bill Duncan, "The Grand Plan for Building an Island Paradise off our Coastline," *Long Beach Press-Telegram*, December 18, 1966.

E. Samuel Pyeatt Menefee, "Republics of the Reefs: Nation-Building on the Continental Shelf and in the World's Oceans," *California Western Journal of International Law* 25, no. 1 (Fall, 1994): 104–05.

6. Ben Sherwood's *The Survivors Club: The Secrets and Science that Could Save Your Life* (New York, Grand Central Publishing, 2009) provides remarkable, hair-raising insights into the minds and genes of survivors of all manner of deadly things.

7. Rob Bender's Web site http://concreteships.org is fascinating. You can still see the hulks of *Jalisco's* nearly indestructible sister ships lined up as a floating breakwater off Powell River in British Columbia and forming the Kiptopeke Breakwater in Lower Chesapeake Bay, Virginia.

8. At least one public record showed a Joe Kirkwood in San Diego who matched the age of Joe Kirkwood Jr. as passing away in 1995. This seems to jibe with a rumor Houtz heard. Yet Kirkwood's Wikipedia page shows no date of death. Kirkwood would now be ninety-one, so of course, it's possible he's still alive. If so, I hope to one day meet him.

CHAPTER 5

1. *Ramapo* fascinated me from the first time I read Sebastian Junger's *The Perfect Storm*. Thanks to Bill Sharp for sharing his copy of "Great Sea Waves" by R.P. Whitemarsh, *U.S. Naval Institute Proceedings* 60, no. 8 (August, 1934). The story cites motion picture footage shot from *Ramapo*. I would appreciate any leads.

2. R.P. Whitemarsh's family contact was made possible through a biography provided by James Allen Knechtmann, at the Naval History and Heritage Command. This led to contact with great-niece Angie Gregos-Swaroop and her mother (and R.P.'s niece) Nancy Whitemarsh Gregos, who provided me with photos and contact information for both James Whitemarsh (R.P.'s nephew) and R.P.'s daughter, Francis "Taffy" Wells.

3. A fascinating analysis of the sinking of the Japanese midget sub by the USS *Ward,* a PT boat under Whitemarsh's command, "The Search for the World War II Japanese Midget Submarine Sunk off Pearl Harbor, Dec. 7, 1941" can be found on the University of Hawaii's Web site (http://www.soest.hawaii.edu/HURL/midget.html).

4. Insights into the science and theory of wave propagation were provided through years of relentless interrogations of Sean Collins for Surfermag.com, *Surfer,* and *The New York Times* (Chris Dixon, "A Site for Real Surfers Catches a Wave," *The New York Times,* June 13, 2002); a telephone interview with Walter

Munk in February 2010; and through two books: Tony Butt and Paul Russell, *Surf Science: An Introduction to Waves for Surfing* (Cornwall, UK: Alison Hodge Publishing, 2002), and Craig B. Smith, *Extreme Waves* (Washington DC: Joseph Henry Press, 2006).

5. 1933 weather stories:

A. "Cold Wave Strikes City Again Tonight; Blizzard in West; Worst Storm in Years Sweeps Middle West," *The New York Times,* February 8, 1933.

B. "Ireland Is Swept by Gale and Snow; Towns Isolated and Shipping Suffers Heavily in Worst Storm of Century," *The New York Times,* February 26, 1933.

C. "Gale-Borne Snow Hits New England," *The New York Times,* February 27, 1933.

D. "6 Die, 20 Missing As Gale Hits Coast; Fishing Boats Lost; Giant Waves Capsize Many Craft, Imperil Hundreds of Fishermen," *The New York Times,* August 21, 1933.

E. "13 Dead, Many Hurt In Pacific Floods; Thousands Are Homeless as Northwest Gets Ray of Hope in Colder Weather," *The New York Times*, December 24, 1933.

F. Paul Bonnifield, *The Dust Bowl: Men, Dirt, and Depression.* (Albuquerque: University of New Mexico Press, 1979).

G. Background on the extreme weather of 1933 was also gleaned from *Monthly Weather Review,* January 1933.

H. 1933 record hurricanes—NOAA (http://www.ncdc.noaa.gov/sotc/tropical-cyclones/2005/13).

I. Oregon weather records—NOAA (http://www.wrh.noaa.gov/pqr/paststorms/index.php).

J. Larry Greenemeier, "Welcome to the Coldest Town on Earth—Oymyakon, Siberia, is bracing for temps as low as minus 90 degrees Fahrenheit," *Scientific American*, December 24, 2008.

6. On the topic of "rogue" or "freak" waves:

A. "Ship-sinking monster waves revealed by ESA satellites," *European Space Agency News*, July 21, 2004 (http://www.esa.int/esaCP/SEMOKQL26WD_index_0.html).

B. Peter Müller, Chris Garrett, and Al Osborne, "Rogue Waves—The Fourteenth 'Aha Huliko'a Hawaiian Winter Workshop," *Oceanography* 18, no. 3 (2005).

C. A fascinating interactive wave machine on PBS's "Savage Seas" Web site (http://www.pbs.org/wnet/savageseas/multimedia/wavemachine.html).

D. Bruce Stutz, "Rogue Waves—The Physics of Pure Hell at Sea," *Discover* Magazine, July 2004.

E. The "Max Wave" project was featured in the BBC Horizon *Freak Wave* series (http://www.bbc.co.uk/science/horizon/2002/freakwave.shtml).

7. To determine the energy in the *Ramapo* wave, the following simple mathematical equation provides a rough but still reasonable estimate: height^2 (meters) x period (seconds) x .5 = kilowatts per square meter of horizontal wave front (if you were able to see how wide the wave stretched across the horizon).

A. In this case: $34(\text{meters})^2$ x 16 x .5 = 9248 kilowatts (a kilowatt being 1000 watts) or 9.2 million watts per square meter.

B. Next, assume this wave presented a front at least two miles wide from end to end—perhaps considerably more. Two miles = 10,560 feet or 3,218 square meters.

C. Thus, 3,218 square meters x 9,200,000 watts = 29,605,600,000 (29.6 billion) watts of energy in this single wave.

8. *Custom of the Sea – A Shocking Tale of Shipwreck, Murder and the Last Taboo,* by Neil Hanson (Wiley Publishing, London, 2000), is the gripping, tragic account of the cannibal crew of the *Mignonette.*

9. The logs of the *Ramapo* are held at the National Archives in Washington DC, and contain entries from Whitemarsh and captain C. B. Mayo.

CHAPTER 6

1. This chapter was made possible through hours of interviews over the course of the past decade (most however, were undertaken when I started research for this book in 2008). Interviewees: Rob Brown, Sean Collins, Bill Sharp, Mike Parsons, Brad Gerlach, Sam George, Walter Munk, Candace Moore, Flippy Hoffman, and Mike Castillo. Other Larry "Flame" Moore material came from interviews Bill Sharp conducted for a 2004 big wave book project to be coauthored by Sharp, Sam George, Steve Hawk, Evan Slater, and myself that was later abandoned—yet Sharp shared it for this book.

2. Regarding Walter Munk's film *Waves Across the Pacific* (1967): Archival video files are available on the Web site of the Scripps Institution of Oceanography (http://libraries.ucsd.edu/locations/sio/ scripps-archives/resources/collections/movie-clips.html).

3. *Surfline*'s video *Making the Call* (2003), available for purchase on Surfline.com, contains fascinating details on what makes Waimea, Maverick's, Todos Santos, Jaws, and the Cortes Bank work.

4. For insight into California surf culture "BG" (Before Gidget), the book *Gidget* by Frederick Kohner (1957, reissued in 2001 by Berkeley Trade) is a must-read.

CHAPTER 7

1. This lede quote comes from Dave Parmenter's seminal "On the Shoulders of Giants," *Surfer*, August 1999.

2. Considerable background was provided by Matt Warshaw, *The Encyclopedia of Surfing* (New York: Harcourt, 2003).

3. The Keaulana quotes in this chapter came from spending a few days with the Keaulana clan in August 2008 for a story in *Men's Journal* magazine. The comments from Greg Noll came by phone and during a visit Twiggy, Greg Long, and I paid him in 2010.

4. Kepelino Keauokalani's description of Hawaiian surfing appear in the book *Kepelino's Traditions of Hawaii* by Kepelino Keauokalani (Honolulu: Bishop Museum Press, 2007).

5. The descriptions of Abner Paki and the recollections of Woody Brown appear on legendarysurfers.com—the incredible Web site of Malcolm Gault-Williams: (http://files.legendarysurfers.com/surf/legends/woody.shtml; http://files.legendarysurfers.com/surf/ legends/ls01_volume1_06.pdf).

6. I interviewed Randy Laine on motorized big wave surfing by telephone in late 2010. I spoke with Herbie Fletcher on the topic back in 2008.

7. Brock Little described his epic Waimea wipeout to Evan Slater for Slater's blog on the Hurley.com Web site and in "Pressure Drop" by Brock Little, *Surfer*, May 1990.

8. Sean Collins's Todos recollections came during an interview at his offices in Huntington Beach in October 2009. I first understood how long period waves worked when I traveled with Sean to cover the Reef@Todos contest in 1999 for Surfermag.com.

9. Rob Brown and Mike Parsons described Mark Foo's death personally in 2009. Evan Slater described Jay Moriarity's wipeout personally in 2009. Ben Marcus's seminal story "Cold Sweat" introduced Maverick's to the world and appears in *Surfer* magazine, June 1992. The film footage of Parsons immediately after Foo's death appears in Mark Matovich's 1994 film, *Monster Maverick's*. There's a clip of the moment on YouTube.

10. The interviews with Laird Hamilton were conducted during work for stories for *Men's Journal* and *The New York Times* in late 2007 and early 2008. The result of some of those interviews appears in a *Men's Journal* story, "Beneath the Waves" (March 2008). The rest of the Laird interviews came about through a piece I was to write for *The New York Times* on the XXL Awards, as a follow-up to the story on the January 2008 Cortes surf mission. A few quick questions with Laird on the premise of the XXLs instead became a full-blown philosophical discussion on big wave surfing in general. I then became buried beneath another story, and the *Times* piece was, unfortunately, never published. Interestingly, this same abandoned story produced the trip to Maverick's where Rob Brown and I learned of Peter Mel's methamphetamine addiction (Chapter 8). I greatly appreciate Laird and Pete Mel's time—and honesty.

11. The "This changes everything" quote from Ben Marcus actually came about watching prerelease footage that would appear in the video *Wakeup Call* with Ben in 1994. In the ensuing years, I would listen to spirited and occasionally angry debates between Sam George, Ben Marcus, Evan Slater, and Steve Hawk over what towsurfing meant and how it should, and should not, be covered in *Surfer*.

12. The video *Wake Up Call* (1995, available on Amazon.com) is a terrific chronicle of the early days of the towsurfing revolution.

CHAPTER 8

1. Everest fatality figures came from the mountain climbing site Alanarnette.com. A 2006 study by the *British Medical Journal* found that overall mortality of 14,138 Everest mountaineers above base camp during the entire eighty-six-year history of climbing was 1.3 percent. J. S. Windsor, P. G. Firth, M. P. Grocott, G. W. Rodway, and H. E. Montgomery, "Mortality on Mount Everest, 1921–2006," *British Medical Journal*, December 2009 (http://www.bmj.com/content/337/bmj.a2654.full).

2. I wrote about motocross casualties for *The New York Times* ("A Motor Sport Takes off Leaving a Trail of Broken Bones," September 2, 2002). In 2000, some 52,000 motocrossers visited American emergency rooms.

3. The interview with Bob Parsons was conducted by telephone in early 2009. I met Jodi Pritchart (Mike's mom) in October 2009 at Mike's boyhood home. Later Cortes interviews with Brad, Mike, and Joe Gerlach were conducted during 2009, while the first Jaws and Cortes interviews happened in 2004.

4. The story on Joe Gerlach: Sandy Treadwell, "Down, Down And Away!" *Sports Illustrated*, March 20, 1972. It's a great insight into thrill-seeking genetics and is also available in SI.com's "Vault" section (http://sportsillustrated.cnn.com/vault/article/magazine/MAG1085900/1/index.htm).

5. Brad Gerlach and Mike Parsons were featured heavily in *Surfer* through the 1980s and '90s in particular. Articles include:

A. Staff, "Testing One Two Three—Brad Gerlach," January 1986.

B. Matt Warshaw, "Talk Show—An Interview with Brad Gerlach," August 1987.

C. Robert Beck, "Todos Santos—¿Quien Es Mas Macho?" June 1988.

D. Matt George, "A For Effort" (lengthy Parsons profile), December 1988.

E. Derek Hynd, "Yankee Domination," January 1992.

F. Jeff Divine, "Size Counts, a High Impact Session at Todos Santos, Baja" (the story of the 1990 "Eddie" and Flame's first "Cortes" swell hitting Todos), June 1990.

G. Steve Barilotti, "Emotional Rescue—The Son of Jumpin' Joe Gerlach Struggles to Save Surfing's Pagan Soul," October 1992.

H. The information on high sensation-seeking came from a telephone interview with Marvin Zuckerman in March 2009, and the book, *Sensation Seeking and Risky Behavior*, by Marvin Zuckerman (Washington DC: American Psychology Association 2008). I would also highly recommend: Florence Williams, "This Is Your Brain on Adventure," *Outside*, April 2009. Read more about Pete Mel and Flea in "Coming Clean – A Searching and Fearless Moral Inventory" by Kimball Taylor, *Surfer*, April 2009, and dying, or not dying in "Death Trip" by Brad Melekian, *Surfer's Journal*, Winter 2011.

CHAPTER 9

1. I first interviewed Dr. Bill O'Reilly in 2003. Follow-up interviews took place in 2007 and 2008. I've nearly driven Sean Collins crazy on the topic of waves for the past decade and spent in-person time with Parsons, Gerlach, Skindog, Mel, Dana Brown, and Rob Brown (no relation) dissecting the January 2001 Cortes session starting as far back as 2003.

2. Stories:

A. Evan Slater, "Project Neptune," *Surfing*, June 2001.

B. Evan Slater, "Bank Job," 2004 (unpublished).

C. Joe Mozingo, "Surfers Catch Monster Waves Off California," *Los Angeles Times*, January 22, 2001.

D. Matt Walker, "The Poseidon Adventure," *Surfline*, January 19, 2001 (http://www.surfline.com/mag/pulse/2001/jan/01_19_cortes.cfm).

E. Evan Slater, "Into Thick Water Part One: Mike Parsons, Brad Gerlach, Peter Mel and Ken 'Skindog' Collins summit Cortes Bank," *Surfline*, January 22, 2001 (http://www.surfline.com/mag/pulse/2001/jan/01_22_bank_one.cfm).

F. Evan Slater, "Into Thick Water: Part Two," *Surfline*, January 22, 2001 (http://www.surfline.com/mag/pulse/2001/jan/01_22_bank_two.cfm).

3. Thanks to Evan Slater for a hilarious comparison of equipment for this surf mission.

JOHN WALLA 1 10'0" Christenson gun • 1 12-foot big-wave leash • 1 pair 3mm booties • 1 3/2mm Billabong fullsuit • 1 bar of cold water Sex Wax

PETER MEL 1 10'3" JC gun • 2 12-foot big-wave leashes • 1 pair 3 mm booties (size 13) • 2 4/3 Quiksilver fullsuits • 1 Quiksilver life vest • 2 7'0" x 16" x 2" JC tow boards (straps included) • 1 30-foot tow-rope • 1 Yamaha XL 1200 WaveRunner • 1 BZ rescue sled • 1 trailer • 2 10-gallon cans of fuel • 1 case of Yamalube oil • 1 roll of duct tape • 1 bag of bungees and tie-downs • 3 oily rags • 1 crescent wrench • 1 hose w/special nozzle • 1 tube of silicone sealant • 1 hammer • 1 vise grip • 1 can engine fogging oil • 1 can Salt Away motor flush • 1 can Engine Stor long-term metal protector • 1 can Corrosion Pro anti-corrosive spray • 1 tube Tri Flow superior lube • 1 can gasoline water absorber • 1 tube performance cable lube • 1 can silicone spray • 1 roll paper towels • 1 funnel • 3 screwdrivers (two flat, one Phillips) • 1 socket wrench • 1 bar cold water Sex Wax

CHAPTER 10

1. In-person interviews were done with Greg, Rusty, and Steve Long, particularly in the fall of 2009. Steve spun a fascinating history of San Clemente and the point break holy land of Trestles. He pointed out, among other things, that Trestles and the San Mateo valley just inland and behind San Clemente were long inhabited by a peaceful tribe called the Ajachemen who farmed, fished, and probably surfed aboard canoes made of the dense reeds that still grow along the rivermouth. Their 10,000-year reign ended after the arrival of Gaspar de Portola and the Spanish in 1769. Portola's men christened each river valley they encountered based on whatever day the next Saint's feast fell. San Mateo Creek honored Saint Mathew while San Onofre was an obscure Egyptian holy man. "Onofre began life as a very pious woman who was to be wed, but she didn't want to marry," said Long. "She went deep into prayer and overnight grew a long beard and men's attributes. She was the first transsexual saint."

2. The videos *What's Really Goin' On* and *What's Really Goin' Wrong* are classic snapshots of not only San Clemente, but American surf culture in the 1990s. They're available from Lost.com.

3. Laird's "Oh My God" wave at Teahupoo appeared on the cover of *Surfer,* February 2001. It also anchors the film *Riding Giants.*

4. To see Mike Parsons's January 7, 2002, Jaws Tow-In World Cup wave that opens the film *Billabong Odyssey*: visit http://tinyurl.com/parsonsjaws and you'll understand why it's so heavily viewed.

5. I interviewed Rush Randle on not competing in the Jaws contest in 2004. Parsons and Gerlach described the Jaws contest to me during an interview in San Clemente in 2004.

6. A great deal of work I would do for *The New York Times* in the coming years—surf-related and otherwise—started with a story called "A Machine Age Clash Among Surfers," February 5, 2002. The story dealt with the unfolding controversy of Jet Ski-assisted surfing through the eyes of Jeff Clark, Peter Mel, Mark Renneker, and Grant Washburn. Some other surf-related *Times* stories included a now tragically timely look at San Onofre, "Atomic Plant Casts a Pall on Paradise," May 12, 2002. An interview with John Parodi, best friend of a victim of the Bali nightclub bombing, appeared in "Lives: The Last Wave," *New York Times Magazine,* November 10, 2002. I described a XXL controversy over a French big wave discovery, "This French New Wave Finds Few U.S. Fans," April 7, 2003. And visited Monterey's deadly giant in "For Daring, Ghost Trees Is Ultimate Thrill Ride," April 3, 2005.

7. A terrific piece on The Billabong Odyssey is by Steve Hawk, "Surf & Destroy," *Outside*, April 2004 (http://outsideonline.com/outside/features/200404/surfing_billabong_odyssey_1.html).

8. I interviewed Dave Kalama for this chapter by phone in early 2011 after reading Susan Casey's *The Wave.* I once interviewed Dave on Maui back in 2001 for *Surfer* on the remarkable "foilboard" surfboard he and Laird had been testing. Kalama then took my future wife and me outrigger canoeing at a rainbow-laced surf break called "Thousand Peaks." My wife still calls the experience among the tops in her life, and said to be sure I thanked Dave again.

9. Results for every XXL contest, including the K2 Big Wave Challenge, were provided by Bill Sharp, along with his "Odyssey" and "Project Neptune" press releases. He was interviewed in *Transworld Business* shortly after getting the go-ahead for the Odyssey, "Billabong Odyssey Offers $500K for 100-Foot Wave," *Transworld Business,* July 10, 2001 (http://business.transworld.net/4676/uncategorized/billabong-odyssey-offers-500k-for-100-foot-wave/).

10. Other stories:

A. Evan Slater, "The Envelope Please . . . Just how will we pick a Swell/XXL grand-prize winner?" *Surfline*, February 2, 2001 (http://ads.surfline.com/mag/pulse/2001/feb/02_22_update.cfm).

B. Matt Walker, "60 (Foot) Minutes—Insider notes from the judges' landmark meeting to decide the XXL Big Wave Awards," *Surfline*, March 3, 2001 (http://el.surfline.com/mag/pulse/2001/mar/03_27_xxl_judging.cfm).

C. Evan Slater, Matt Walker, Marcus Sanders, and Steve Hawk, "Mike Parsons wins the Swell-Surfline XXL Big Wave Awards for his 66-foot bomb at Cortes Bank," *Surfline*, March 3, 2001 (http://ads.surfline.com/mag/pulse/2001/mar/03_30_ceremony.cfm).

D. Mike Parsons as told to Chris Mauro, "Run Silent Run Deep: A Secret Mission to Cortes Bank Kicks off the 2003/2004 Big Wave Season," *Surfer,* April 2004.

E. Pete Thomas, "Remote Surf Spot Draws Mob," *Los Angeles Times,* January 20, 2004.

F. Marcus Sanders, "Cashing In: Six Surfers Tackle Giant Surf at the Cortes Bank," *Surfline,* March 10, 2005 (http://www.surfline.com/surfnews/2005_03_10/cashing_in_story.cfm).

G. Brad Gerlach, "To Hell and Back," *Surfline*, Feb 15, 2005 (http://www.surfline.com/surfnews/2006_02_15_cortes.cfm).

H. Terry McCarthy, "When the Surf's Way Up," *Time*, July 19, 2004 (http://www.time.com/time/magazine/article/0,9171,662800-2,00.html).

I. Susan Casey, "The Jaws Paradigm," *Sports Illustrated*, August 7, 2006 (http://sportsillustrated.cnn.com/vault/article/magazine/MAG1105277/index.htm).

J. Motoko Rich, "Author and Surfer to Share in Proceeds of Waves Book," *The New York Times*, May 9, 2007 (http://www.nytimes.com/2007/05/09/books/09surf.html).

CHAPTER 11

1. The first round of interviews for this chapter came as I was working on "Beneath the Waves," *Men's Journal,* March 2008. After Peter Davi died, I spent December 23, 2007, with Pete's eighteen-year-old son Jake and Pete's buddy Anthony Ruffo. Jake was unaware at the time that a toxicology report would show methamphetamine in Pete's blood. I greatly appreciate Jake and Ruffo's help with the story and hope that they feel that while I couldn't overlook the report, I also didn't overlook the fact that Pete was a fiercely intelligent, pioneering, and highly respected waterman.

2. A stellar play-by-play of the December 2007 swell appears in "Chasing Giants" by Greg Long and Mark Healey, *Surfing*, March 2008.

3. By January 6, 2008, I was gawking at stupefying video footage alongside Greg Long, Twiggy, Snips, and Gerlach at Mike's home in the wake of the January 5 mission to Cortes. Coverage began with:

A. "Eye of the Storm. Gerlach, Snips, Greg Long and Twiggy Score Huge at Cortes Bank" by Mike Parsons, Surfline, January 8, 2008. (http://www.surfline.com/surfnews/photo_bamp_html .cfm?id=13060&pic=2).

B. "Surfers Defy Giant Waves Awakened by Storm," by Chris Dixon, *The New York Times*, January 9, 2008 (http://www.nytimes.com/2008/01/09/sports/othersports/09surf.html).

C. "The Next Sir Edmund Hillarys: Riders on the Storm" by Lawrence Downes, appearing on The Board: A Blog by the Editorial Writers of *The New York Times,* January 11, 2008 (http://theboard .blogs.nytimes.com/2008/01/11/the-next-sir-edmund-hillarys-riders-on-the-storm/).

4. The funniest takes on the Cortes Mission came from Surfermag.com's forum: "Parson's XXXXXXXXL" (http://forum.surfermag.com/forum/showflat.php?Number=1299798).

5. The story of the great storm of 2008 came through a interviews with Leanne Lusk, command duty officer for the Coast Guard's eleventh district, and Dr. Warren Blier, Chief Science and Operations Officer at NOAA in San Francisco.

6. Other sources:

A. NOAA meteorologist Scott Hazen Mueller provided day-by-day updates of the storm's progress on the National Weather Service's San Francisco office Web site. (http://groups.google.com/group/ ba.weather/search?q=dec+31+2007+discussion&start=0&scoring=d&)

B. Jeffrey Lewitsky, Arlena Moses, and Robert Ruehl, "January 3–6, 2008 Storm Summary/Case Study," NWS WFO Eureka, California, 2008 (www.wrh.noaa.gov/eka/local_studies/2008-Jan-3-6 .pdf).

C. Jesse McKinley, "Ferocious Storm Punishes Northern California," *The New York Times*, January 5, 2008.

D. Gary Klein, "Sausalito Diver Dies After Fall into Frigid Bay," *Marin Independent Journal*, January 7, 2008.

E. "Public Information Statement," National Weather Service, San Joaquin Valley, California, January 9, 2008 (www.wrh.noaa.gov/hnx/events/Jan042008rain.txt).

F. "Preliminary Local Storm Report . . . Summary," National Weather Service, San Joaquin Valley, California, January 5, 2008 (www.wrh.noaa.gov/hnx/events/Jan042008wind.txt).

G. Douglas Le Comte, "Weather Highlights 2008," *Weatherwise*, March–April 2009.

H. "Maximum Height Of Extreme Waves Up Dramatically In Pacific Northwest," Oregon State University, January 25, 2010 (http://oregonstate.edu/ua/ncs/archives/2010/jan/maximum-height-extreme-waves-dramatically-pacific-northwest).

7. "January 4–5, 2008 Winter Storm Summary," National Weather Service, Hanford, Oregon, January 2008 (http://www.wrh.noaa.gov/hnx/jan042008winterstorm.php).

8. "2008 New Years Storms," National Weather Service, San Francisco/Monterey Bay, January 2008 (http://www.wrh.noaa.gov/mtr/jan4_storm.php).

9. "Top Weather Stories for the Decade, 2000–2009," National Weather Service, Reno, Nevada, January 2009 (www.wrh.noaa.gov/rev/climate/decade00.php).

10. David R. Baker, "PG&E's Blacked-Out Users During Storm Might Be 10-Year Record," *San Francisco Chronicle*, January 8, 2008.

11. "Storms kill three in western US," BBC, (http://news.bbc.co.uk/2/hi/americas/7173995.stm), January 6, 2008.

12. "Man Killed, Seven Injured on Icy Mountain Highway," KVAL TV (http://www.katu.com/news/13486032.html), January 6, 2008.

In the wake of Larry "Flame" Moore's death, Candace and Larry's sister Celeste have formed the Follow the Light Foundation, which bestows grants on budding surf lensmen (http://followthelightfoundation .org). I could not have properly told Flame's part of this tale had his wife, Candace, not graciously invited me into her home and related a very difficult story of love and loss. Without Flame's pioneering efforts, this book would have never been written, and I can only hope Flame would think it a worthy read.

Late into my work on this book, my own father was diagnosed with late-stage lung cancer. I'm thankful he was able to read an early copy of this book before he passed. I'd like to dedicate the pages you've just read to my dear old dad, Richard Jobie Dixon, July 17, 1943–July 3, 2011.